Phase Change Memory

From Devices to Systems

Synthesis Lectures on Computer Architecture

Editor
Mark D. Hill, *University of Wisconsin, Madison*

Synthesis Lectures on Computer Architecture publishes 50- to 100-page publications on topics pertaining to the science and art of designing, analyzing, selecting and interconnecting hardware components to create computers that meet functional, performance and cost goals. The scope will largely follow the purview of premier computer architecture conferences, such as ISCA, HPCA, MICRO, and ASPLOS.

Phase Change Memory: From Devices to Systems

Moinuddin K. Qureshi, Sudhanva Gurumurthi, and Bipin Rajendran

ISBN: 978-3-031-00607-4   paperback
ISBN: 978-3-031-01735-3   ebook

DOI 10.1007/978-3-031-01735-3

A Publication in the Springer series
SYNTHESIS LECTURES ON ADVANCES IN AUTOMOTIVE TECHNOLOGY

Lecture #18
Series Editor: Mark D. Hill, *University of Wisconsin, Madison*
Series ISSN
Synthesis Lectures on Computer Architecture
Print 1935-3235   Electronic 1935-3243

Phase Change Memory

From Devices to Systems

Moinuddin K. Qureshi
Georgia Institute of Technology

Sudhanva Gurumurthi
University of Virginia

Bipin Rajendran
IBM Research

SYNTHESIS LECTURES ON COMPUTER ARCHITECTURE #18

ABSTRACT

As conventional memory technologies such as DRAM and Flash run into scaling challenges, architects and system designers are forced to look at alternative technologies for building future computer systems. This synthesis lecture begins by listing the requirements for a next generation memory technology and briefly surveys the landscape of novel non-volatile memories. Among these, Phase Change Memory (PCM) is emerging as a leading contender, and the authors discuss the material, device, and circuit advances underlying this exciting technology. The lecture then describes architectural solutions to enable PCM for main memories. Finally, the authors explore the impact of such byte-addressable non-volatile memories on future storage and system designs.

KEYWORDS

phase change memory, non-volatile memory, storage, disks, systems

Contents

Preface

The memory hierarchy plays a pivotal role in determining the overall performance of a computer system. Many applications, especially those in the server space, have large working sets and hence require high memory capacity. The demand for greater memory capacity is expected to continue into the future, even for commodity systems [111]. Several applications are also I/O intensive and hence require a high-performance storage system. Furthermore, it is important to minimize the speed gap between the processor, memory, and the storage system in order to achieve high performance. Finally, power consumption is an important criterion in designing and deploying computer systems and affects the overall electricity usage of the computing infrastructure and also cooling costs, both of individual machine components and also at the data center scale.

The memory hierarchy of modern computers is designed such that the layers closer to the processor cores, such as the caches, provide low latency access whereas those layers farther away (e.g., main memory) provide higher capacity, albeit with longer access times. Moreover, the lowest layer of the memory hierarchy - storage - has also provided an additional property called *non-volatility*. A non-volatile medium is that which allows data to be read and written, but also for the written data to be retained for extended periods of time without the need for an external power supply to preserve the state. Static RAM (SRAM), Dynamic RAM (DRAM) and rotating disks have served as the bedrock technologies for designing processor caches, main memory, and storage for many years. However, continued use of these technologies poses several challenges. While SRAM provides low latency, it suffers from excessive power consumption (especially leakage power) and low density. DRAM provides high density, but again suffers from high power consumption and scalability problems [8]. While hard disk drives provide low cost-per-Gigabyte of storage, they suffer from high access latencies (several orders of magnitude higher than main memory) and also consume a significant amount of power. While replacing rotating disks with Flash memory, in the form of Solid-State Disks (SSDs), can provide a large performance boost for storage, there is still a wide gap in the performance between main memory and storage that needs to be narrowed.

One way to address the density, power, performance, and scalability problems of the traditional memory and storage technologies is to use a *Non-Volatile Memory (NVM)*. Non-Volatile Memories are a set of memory technologies that provide the non-volatility property. While there are several promising NVM technologies that are being actively explored by academia and industry, the one technology that is closest to large-scale production is *Phase Change Memory (PCM)*. PCM uses phase-change material that can be in one of two states - crystalline or amorphous - each of which has a distinct resistance that can be attributed to the state of a bit. Moreover, PCM provides the ability to store multiple bits of data per memory cell and the latency of PCM-based main memory is closer to that of DRAM. While PCM provides several benefits, it also poses certain key challenges, such as its limited write endurance and high write latency, that need to be addressed before this memory technology can be leveraged to build high-performance computer systems. The non-volatility property of PCM (and NVMs, in general) provide new opportunities and challenges for both computer architecture and the design of software systems.

In this Synthesis Lecture, we provide a detailed overview of PCM and its use at three levels of system design: *devices, architecture, and systems*. We begin with an overview of the key physics that governs non-volatile memory, discuss several NVM technologies, and then provide a detailed exposition of PCM. We then discuss how PCM can be leveraged to architect main memory systems, discussing issues related to the latency, power, and reliability. We then discuss storage and systems issues related to the use of PCM and PCM-like Non-Volatile Memories.

The organization of this Synthesis Lecture is as follows:

- Chapter 1 discusses the requirements for a next generation memory technology and briefly surveys the leading candidates that vie to fulfill these requirements. It then provides an overview of PCM material, device and circuit level innovations and discusses the outstanding challenges PCM has to overcome to enable their use in commercial applications.

- Chapter 2 discusses some of the recent proposals to leverage PCM in future main memory systems - particularly hybrid architecture schemes that use both PCM and DRAM as well as PCM-only approaches, and the related optimizations.

- Chapter 3 describes write cancellation and write pausing as a means to reduce the read latency increased caused by slow writes in PCM.

- Chapter 4 describes an efficient technique to perform wear leveling at line granularity while keeping the storage, latency, and extraneous write overhead to negligible level.

- Chapter 5 introduces the security problem in lifetime limited memories and provides an overview of recent security-aware wear leveling algorithms along with circuits that can detect such attacks at runtime.

- Chapter 6 describes efficient hard error correction for phase change memories including both hardware-software proposals and hardware-only proposals.

- Chapter 7 discusses systems level issues in using PCM and PCM-like byte-addressable NVMs. This chapter first discusses the use of the emerging NVMs in storage system design and presents a case study of a prototype PCM-based SSD. The second part of this chapter considers a broader set of systems issues from recent literature that examine the impact of the non-volatility property of NVMs to design software systems, memory interfaces, and optimize performance and energy-efficiency.

This Synthesis Lecture is the outcome of a series of tutorials developed and presented by the authors at the 2010 IEEE International Symposium on High-Performance Computer Architecture

(HPCA), the 2010 IEEE/ACM International Symposium on Microarchitecture (MICRO), and the 2011 Non-Volatile Memories Workshop (NVMW). As was the case for the tutorials, this Synthesis Lecture is intended for a broad audience: graduate students, academic and industry researchers, and practitioners. We believe that this lecture will serve as the starting point for the reader to learn about the devices, architecture, and systems concepts and ideas related to this exciting memory technology.

Moinuddin K. Qureshi, Sudhanva Gurumurthi, and Bipin Rajendran
October 2011

CHAPTER 1

Next Generation Memory Technologies

1.1   INTRODUCTION

The urge to record and transmit information has been one of the primary driving forces that has fuelled the progress of human civilization. Long before the advent of paper and modern printing, human beings have been recording information on 'non-volatile memory substrates' – the 'Lebombo bones' discovered within the Border cave in Swaziland still retains the information (29 notches representing the duration of a lunar month) that was engraved on them some 37, 000 years ago [144].

During the last fifty years, we have made tremendous strides in documenting and retrieving information, thanks to the advances in semiconductor and magnetic memory devices. At the time of writing, all the information contained in the U.S. Library of Congress's web archives can be stored in approximately 100 hard-disk drives, at a cost of about US\$ 5, 000 [134]. The ability to continually double the number of transistors approximately in every 18 months by shrinking their critical dimensions – known as Moore's Law scaling [125] – has also resulted in impressive progresses in computing capabilities. For instance, today's supercomputers can complete more than 10^{15} floating point operations per second (FLOPS) [53].

However, sustaining this rate of performance enhancements in computation, memory or storage into the next decade will become extremely challenging, due to limitations set by fundamental physics as well as economic factors [26, 87]. At the same time, our appetite for faster and cheaper computational capabilities is only growing with time. It is becoming evident that the realization of 'exa-scale' (10^{18} FLOPS) computational systems will require fundamental breakthroughs in memory capacity, speed of data access and reliability [201]. New ideas, materials and fabrication techniques will be necessary as it is unclear if conventional device scaling can provide the technology platforms necessary for these future systems. Novel materials such as chalcogenide alloys [94] and nano-scale fabrication techniques such as self-assembly [6] and nano-imprint lithography [32] hold promise to address this challenge.

In this chapter, we will first describe some of the desirable characteristics of a next generation memory technology. We will then survey the key non-volatile memory candidates starting with Flash memory. The number of emerging memory candidates is quite vast, spanning a large number of material systems such as oxides, chalcogenides, organics and polymers, just to mention a few. In order to provide a concise and relevant exposition that could be useful to device and material physicists as well as computer architects and system designers, we will restrict our study to those technologies that have already demonstrated Mbit-scale memory chips, which is a typical metric for technology maturity. The bulk of the chapter will be spent in explaining the fundamental physics of Phase Change Memory (PCM) materials and devices, including multi-level cell (MLC) programming. We will conclude the chapter by identifying the major reliability challenges associated with PCM and some potential solutions to address them.

1.2   NEXT GENERATION MEMORY TECHNOLOGIES – A DESIDERATA

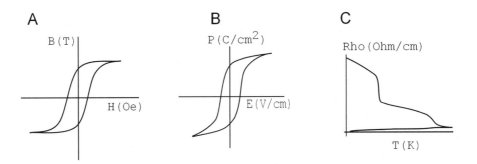

Figure 1.1: Hysteresis curves of (A) ferromagnetic, (B) ferroelectric and (C) chalcogenide memory materials showing more than one stable physical state. When configured into memory devices, these different physical states result in a measurable difference in an electrical quantity such as its effective resistance or capacitance.

Any material that exhibits some form of hysteresis in one of its electrically probable property is a candidate for realizing a memory device. Typically, this means that the material exist in more than one stable physical state, with a measurable difference in an electrical quantity such as its effective resistance or capacitance; with transition between such states mediated by external energy input in the form of voltage or current pulses (Figure 1.1). However, to become a cost-effective, viable technology, a set of stringent and exacting requirements have to be satisfied in addition to this basic criterion. We begin by listing some of the high-level features that are desirable in a next-generation memory technology:

(a) Non-volatility – the ability to retain stored data for an extended period of time, in the absence of external power.

(b) No moving parts – preference for a complete solid-state solution to enable integration with lean, mobile computing platforms of the future.

(c) Low cost – memory market, being extremely cost sensitive, makes it inevitable that any new device have the potential to scale and achieve a target of less than US$ 1 per Giga-Byte (GB) of memory[1], and

(d) Independent of transistor scaling – It is preferable for the fundamental data storage mechanism not to rely on transistor operation, due to the difficulties anticipated in scaling the critical dimensions and fundamental operating characteristics of complementary metal-oxide-semiconductor (CMOS) transistors.

Table 1.1: Specifications for a next generation memory technology.

Metric	Specification
Program/access time	$< 50 - 100$ ns
Programming energy/bit	$< 1 - 10$ pJ
Cell size	$< 10F^2$
Number of bits per physical device	$2 - 4$
Data retention	> 10 years at $85\,°C$
Repeated programmability	$> 10^8$ cycles
CMOS compatibility (processing temperature)	$< 450\,°C$

In addition to these features, an emerging memory technology should also satisfy certain specific requirements to meet the demands of the computational systems of the future, as listed in Table 1.1. The first criterion in this list relates to read and write latency of the memory device – it should be possible to complete the read and write operation at the device level in the range of $10 - 100$ ns, so as to maintain reasonable data transfer rates between the memory and microprocessor blocks in a modern computational system [158]. The second specification for a good memory device candidate is set by the operating power of the memory chip, which is dependent on the total energy spent in programming the device. This translates to a write current in the range of $10 - 100\mu A$ for a device programmed with a voltage pulse of amplitude 1V and pulse width of approximately 100 ns.

The next two criteria are related to the achievable capacity of the memory system. The physical area occupied by the memory cell should be less than $10F^2$, where F is defined as the smallest lithographic dimension in a given technology node. Note that the smallest possible area for a lithographically defined memory cell in a uniform array of cells is $2F \times 2F = 4F^2$, since 1F is necessary to pattern the memory element itself and another 1F is necessary to pattern the minimum possible spatial separation between two adjacent devices [26]. As a point of reference, typical dynamic random-access memory (DRAM) cell size is in the range of $4 - 8F^2$ [177] while the area of a typical static random-access memory (SRAM) cell size is in the range of $125 - 200F^2$ [79]. So, the requirement that memory cell area of an emerging contender should be less than $10F^2$ is really a comparative statement with respect to the achievable density and cost of a DRAM chip. Further, each physical memory cell in technologies such as NOR and NAND flash could be programmed to distinct programmed states, thus storing multiple information bits in each device ($\log_2$[Number

---

[1]At the time of writing, DRAM costs about US$ 10 per GB, NAND flash is approaching US$ 1 per GB and hard-disk drive is approaching US$ 0.1 per GB.

of resistance levels] = number of bits/cell). This feature, referred to as multi-level cell (MLC) pro-grammability, is desirable as it allows to approximately double or quadruple the memory capacity without proportionately increasing the physical real-estate devoted to memory.

It should also be noted that typical memory systems are configured in a hierarchical manner, comprising of two-dimensional arrays, with memory cells connected at the intersection of perpendic-ular metal lines, called 'bit-lines' and 'word-lines' (Figure 1.2). In the periphery of such regular arrays of memory devices are CMOS logic circuits that allow controlled access to the memory cells for programming and reading the data stored in them. Also, in order to maintain good signal-to-noise ratio during sensing and prevent parasitic disturb of neighbouring devices during programming, es-pecially in large memory arrays, a switch with high on-off ratio (typically $> 10^6$) such as a rectifying diode, bipolar junction transistor (BJT) or CMOS field effect transistor (FET) is incorporated in series with the memory device [51]. Further, it is also typical to provide extra memory devices and associated decoding circuitry on each word-line for redundancy and error correction. The capacity of a memory array is thus dependent not only on the area of the fundamental memory cell and its MLC capability, but also on the area of the peripheral circuits and the extra bits that are necessary to main-tain the reliability of stored data. A memory designer has to pay close attention to all these aspects in arriving at an optimal design solution. It is also worth mentioning that there are other exciting fabrication ideas that are pursued such as three-dimensional integration [162] and sub-lithographic cross-bar arrays [50] to enhance memory capacity beyond what is possible by conventional methods.

The next two specifications in Table 1.1 are related to the reliability of the memory cell over prolonged use. There are two aspects to this: data retention and endurance. The first specification states that it should be possible to maintain the operating characteristics of the memory array even after subjecting it to 85 °C for a period of 10 years, which is typically the qualifying specification for flash memories [23]. Since the rate of data loss for memory technologies is typically an exponentially activated process, data retention measurements are typically conducted at elevated temperatures (in the range of $100 - 200$ °C), from which the retention times at 85 °C can be obtained by extrapo-lation [133]. Endurance specification relates to the number of program/erase cycles each individual device in the array can be subjected to before failing to operate within the specified windows of op-eration. For DRAM, this is believed to be in excess of 10^{15} cycles, while the endurance specification of flash memories have been drastically decreasing with scaling (now less than 10^4) [21]. Seen in this light, meeting the endurance specification target of 10^8 cycles is quite aggressive for a nanoscale next generation memory technology. It should also be noted that both the endurance and retention targets are harder to meet in MLC technologies compared to their 1-bit counterparts.

It is also of critical importance that the fabrication steps involved in the integration of novel memory candidates into an integrated chip are compatible with conventional CMOS processing which is used to build the peripheral control circuits of the chip. Typically, memory elements are integrated in the process flow after the completion of the fabrication steps for the CMOS devices, and this sets an upper limit to the thermal budget that the substrate can be subjected to, primarily to maintain the reliability of the semiconductor junctions and the low resistive silicide contacts of

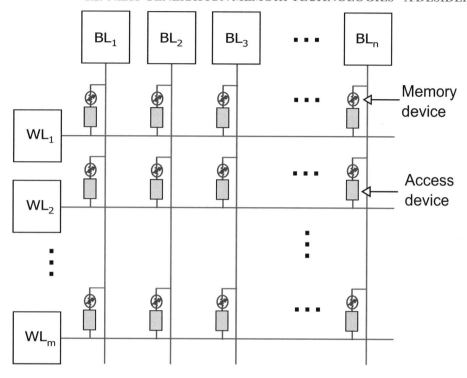

Figure 1.2: Typical memory systems are configured in a hierarchical manner, comprising of two-dimensional arrays, with memory cells connected at the intersection of perpendicular metal lines, called 'bit-lines' and 'word-lines'. A generic memory cell is represented in this figure by a access device (such as diode, FET, BJT, etc.) in series with a programmable memory device that exhibits hysteresis. In the periphery of such regular arrays of memory, devices are CMOS logic circuits that allow controlled access to the memory cells for programming and reading the data stored in them.

the transistors. Hence, it is ideal if the maximum temperature necessary for the integration of the memory device can be limited to 450 °C [162].

The physical mechanisms underlying many of these desirable characteristics are intimately coupled with each other and careful engineering is necessary to trade-off one feature (for instance, retention time) to improve another (e.g., speed and programming energy) [192]. Hence, the development and integration of a memory device meeting all these exacting, and often competing requirements can be a challenging task and calls for careful materials and device optimization.

1.3   OVERVIEW OF FLASH MEMORY AND OTHER LEADING CONTENDERS

Having set the expectations, we now proceed to understand the basic characteristics of some of the current leading contenders in the race for the next generation memory technology. We begin with a discussion on flash memory to understand why in spite of being mature and well-established, the search is on for an alternative technology.

1.3.1   FLASH MEMORY

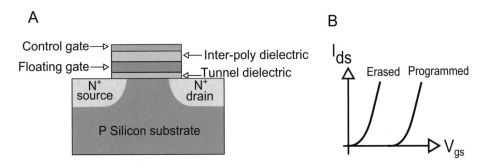

Figure 1.3: (A) Structure of flash memory cell showing the floating gate surrounded by dielectric material. (B) I-V characteristics of a flash memory cell in the programmed state and erased state – trapped charge carriers within the floating gate shift the effective threshold voltage of the transistor to the right. (Adapted from [23]).

The central idea behind the Flash memory cell is simple – it is a transistor with an electrically programmable threshold voltage (For a good review, see [19, 23, 95, 143]). This is achieved by introducing an additional conducting layer in the gate stack of the conventional field effect transistor (FET) called the floating gate, completely surrounded by dielectrics (Figure 1.3A). The memory cell is programmed by applying bias voltages to the control gate and the drain, which causes tunnelling and trapping of charges in the floating gate, resulting in a shift of the effective threshold voltage of the device (Figure 1.3B). In the absence of external fields, the trapped charges within the floating gate are quantum mechanically confined, preventing them from escaping the potential barrier set by the large band-gap of the surrounding dielectric. The data stored in the memory cell is read by applying a small voltage (< 1 V) at the drain and measuring the current through the transistor. The dielectric layers have to be thick enough to prevent charge leakage after programming but thin enough to allow program/erase within reasonable amounts of time, while maintaining good coupling ratio (defined as the ratio of the control gate to floating gate capacitance and the overall floating gate capacitance). Poly-silicon is typically used for forming the gate layers, while silicon di-oxide (SiO_2), ONO films (essentially a triple layered structure, comprising of a layer of silicon nitride (Si_3N_4) deposited between two SiO_2 layers) or high-k dielectric materials could be used for forming the dielectric layers. For the 22 nm technology node, the International Technology Roadmap for Semiconductors (ITRS) projects that the effective tunnel dielectric thickness should be in the range of $6 - 8$ nm and that of the inter-poly dielectric should be in the range of $9 - 10$ nm [69].

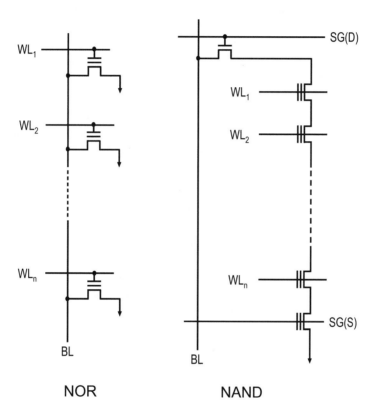

Figure 1.4: In NOR flash array, each memory cell is connected to the common drain contact making each of them directly accessible through the bit-line and word-line. In NAND flash array, multiple memory cells are configured in series, eliminating the source and drain contacts between devices, resulting in a more compact array (Adapted from [95]).

Flash memory can be broadly classified into two categories – NOR flash and NAND flash memory, based on the organization of the memory array (Figure 1.4). In NOR flash array, each memory cell is connected to the common drain contact making each of them directly accessible through the bit-line and word-line. In NAND flash array, multiple memory cells are configured in series, eliminating the source and drain contacts between devices, resulting in a more compact array; the trade-off associated with this dense packing is that read speed for NAND flash (< 25 MBps) is slower than that of NOR flash ($\sim$ 100 MBps). At 65 nm node, typical cell sizes demonstrated for NOR and NAND flash memory cells (accounting for MLC) are $\sim 5F^2$ and $\sim 2F^2$, respectively [95]. At the time of writing, the most advanced NAND flash memory chip in the literature was a 64 Gb chip employing 3 bits/cell technology in 20 nm CMOS technology, with 64 memory cells connected in series within a block [141].

NOR flash memory is programmed (defined as the operation to increase the threshold voltage of the FET) using a phenomenon called channel hot-electron injection (CHEI), where at sufficiently large drain bias (4 − 6 V) and gate bias (8 − 11 V) conditions, minority carriers in the channel accelerated by the drain field undergo impact ionization, gain enough energy to overcome the potential barrier of the dielectric and end up trapped within the floating gate. NOR flash memory erasure depends on a phenomenon called Fowler-Nordheim (FN) tunnelling, which is a quantum mechanical process by which electrons 'tunnel' through the triangular potential barrier created by the applied field (> 10 MV/cm). The typical cell gate bias for erase is −10 V at the control gate and 5 V at the source, while the drain and bulk terminals are kept at ground. CHEI programming is accomplished in few microseconds, while the FN erasure takes few milliseconds. Hence, a collection of cells in a string/block are erased in parallel. To read the cell, the control gate is biased at about 5 V and the drain to about 1 V and the current through the cell ($10 - 50\mu A$) monitored to determine the state of the cell.

Both programming and erasing of NAND flash memory is based on FN tunnelling, with voltages in the range of 15 − 20 V. The control gate is biased to high voltage for programming, while the substrate is biased to the high voltage for erasing. As the programming currents are in the nA range, multiple cells can be programmed in parallel, resulting in a higher write bandwidth (> 5 MBps) for NAND flash compared to NOR flash (< 0.5 MBps) technology. Iterative programming techniques are often employed to tighten the distributions in threshold voltages of the devices in an array, and is necessary if MLC capability is required. The general strategy is to apply voltage pulses with increasing amplitudes interleaved with verify-read cycles to assess if the cell has achieved its target threshold voltage. For performing the read operation, the control gate of the selected device is held at ground, while the control gates of all the other devices on the BL are biased at $\sim$ 5V. The state of the selected device is then determined by a sense amplifier on the BL that measures the current in the line while the BL voltage is clamped to a small voltage (< 1 V).

It should be now clear that the primary objective in the design of flash memories, especially NAND flash, is to obtain the highest possible memory density. In fact, flash memory scaling has been one of the primary driving forces behind Moore's Law scaling, which in turn enabled its per bit cost to drop by a factor of more than 8, 000 in the 20 year period from 1987 − 2007 [95, 145]. However, this amazing reduction in cost comes at the cost of relatively poor read and write performance in comparison with DRAM or SRAM as well as a reduction in the maximum number of program/erase cycles with technology scaling. In fact, the nominal endurance specification for NAND flash memory has dropped from 10^5 cycles in 90 nm to 10^4 cycles in 50 nm technology [21]. Beyond 22 nm node, the nominal endurance specification is not expected to exceed 10^3 cycles. This brings us to the main limitations of flash memory technology and why scaled flash memory can not satisfy the memory requirements of a next generation computational system.

A number of path breaking innovations in device engineering, lithography and system design has enabled flash memory scaling over the years. Some of these worth mentioning that are already incorporated into devices or that are being actively studied for potential deployment are advanced multi-layered gate stacks such as SONOS [58], TANOS [101], high-k dielectrics [193], nano-crystals [126], double patterning [59], MLC programming (3 bits per cell is already demonstrated) [141] and a 3-dimensional channel structure for future Tbit storage [77]. However, we are now reaching a point where many of the materials and device requirements to continue this scaling trend are meeting fundamental physical road-blocks or are becoming economically prohibitive [85].

As the critical dimensions scale down, the numbers of electrons that are stored at the floating gate per level is decreasing (for instance, approximately 100 electrons are stored at the 30 nm node corresponding to a threshold voltage shift of +6 V), to the extent that the loss of few electrons due to leakage can alter the state of the memory cell [85]. Further, it is become extremely challenging to maintain the aspect ratio of the gate stack needed to maintain a good coupling ratio, while keeping the cross-talk between adjacent cells manageable [23]. Since large electric fields are necessary to induce tunnelling of carriers over the potential barriers in excess of 3 eV, the quality of the dielectrics degrade quickly with cycling, especially as the dielectric thickness scale to the sub-10 nm regime [85]. The drastic reduction in the number of program/erase cycles and the limitations related to read/write bandwidth and random access are proving to be big obstacles in the path of deployment of flash memories in large enterprise systems. This has prompted the semiconductor materials and devices community to search for alternative candidates that can satisfy the requirements of future computational systems.

1.3.2   FERRO-ELECTRIC RAM

Ferro-electric RAM or FeRAM was one of the earliest contenders to vie for the position of a storage class memory technology [25]; the earliest proposal was to use the "snapping dipoles of ferro-electric materials" to store binary data [24, 153]. Polarization refers to the displacement of electronic charges in materials due to the application of external electric fields. In conventional dielectrics, the induced polarization vanishes when the external field is removed. In ferro-electric materials such as barium titanate ($BaTiO_3$), lead zirconate titanate ($PbZr_xTi_{1-x}O_3$), strontium bismuth tantalite ($SrBi_2Ta_2O_9$), there are two stable states of polarization corresponding to the two stable configuration of ions within the unit cell of the crystal lattice. This polarization (called remnant polarization, P_r) does not vanish when the external field is removed (Figure 1.5 A).

The memory element consists of a ferro-electric capacitor made by sandwiching the ferro-electric material between metallic electrodes made of Pt, Ir, or oxides of transition metals such as RuO_2 and IrO_2. An access device (typically FET) is also needed to prevent cross-talk during programming and reading, configured either as a series element with the capacitor (1T-1C cell) [55, 89] (Figure 1.5 B) or as a string of FETs and capacitors connected in parallel, with the BL

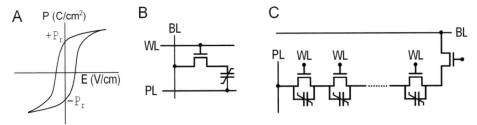

Figure 1.5: (A) Ferro-electric materials have two stable states of remnant polarization (in the range of $10 - 30\mu C/cm^2$), which can be used to encode binary data. Electric field required for programming is in the range of $0.5 - 1$ MV/cm. The memory cell consists of a ferro-electric capacitor, with an access device (typically FET), configured either as (B) a series element with the capacitor (1T-1C cell) or (C) as a string of FETs and capacitors connected in parallel (chain FeRAM). (Adapted from [68, 89, 197]).

contact shared with multiple cells (chain FeRAM) [197] (Figure 1.5 C). There are also proposals to integrate ferroelectric materials as the gate insulator of transistors to obtain a compact memory cell, though large chip-scale demonstrations utilizing this structure have not been reported [117].

The memory cell is programmed by applying a voltage pulses to the bit-line (BL) to write **1** and to the plate-line (PL) to write **0**. In the case of the 1T-1C cell, the word-line (WL) of the selected device is turned on to access the selected device, while the WL of the unselected devices is turned on in the case of chain FeRAM to bypass the unselected devices. Attainable cell sizes with these configurations are in the range of $4 - 15F^2$. The main attractiveness of this technology is that the read and programming pulses can be lesser than 50 ns and it is possible to get endurance typically about 10^{12} cycles.

The main limitation of FeRAM is that the read operation is destructive – in the typical measurement scheme, a voltage pulse is applied to the PL that takes the capacitor to one extreme of its hysteresis loop. If the remnant polarization in the ferroelectric capacitor is in a direction corresponding to the read voltage, a relatively small current flows through the ferroelectric capacitor. If the remnant polarization initially opposes the read voltage, the read voltage flips the direction of the remnant polarization, resulting in a relatively large current. As a result of the application of the read pulse, the capacitor always ends up in the same final state independent of the original stored data. Therefore, to maintain data integrity, each read operation has to be immediately followed by a write operation, so that the read data can written back into the memory cell. This write-after-read translates into a large endurance requirement for the memory cell ($> 10^{16}$ cycles) [76].

Another issue that confronts FeRAM is its scalability. Typical values for the remnant polarization for the commonly used ferroelectric materials is in the range of $10 - 30\mu C/cm^2$, which translates to an effective charge of ~ 1000 electrons for a capacitor with an area of 20nm $\times 20$ nm. This means that with conventional area scaling, the sense margin will continue to decrease, unless 3−dimensional structures are not employed. A further challenge is that the temperature necessary for fabrication of the ferroelectric capacitor is in the range of $600 - 700\,°C$ due to the high crystallization temperature of these materials [41].

At the time of writing, the largest FeRAM memory chip demonstrated in literature is a 128 Mbit chain-FeRAM from Toshiba Corporation, built in 0.13μm process technology [187]. The memory cell size in this demonstration was 0.45μm$\times0.56\mu$m$= 0.252\mu$m^2, with a cell charge of about ~ 13 fC and a cell signal of ~ 100 mV. The chip achieved a data-rate of 800 Mb/s/pin in read/write operation and 10^{13} read/write burst cycles. FeRAM products are used in niche applications such as RFID tags [128] and SONY Playstation 2 [37].

1.3.3   MAGNETIC & SPIN-TORQUE TRANSFER RAM

The origins of Magnetic RAM or MRAM can be traced back to the invention of ferrite core memory in the early 1950s, where binary information was encoded in the clockwise/anti-clockwise magnetic polarity of the donut shaped ferrite cores [127]. The cores were programmed by the vector sum of the magnetic fields created by the currents flowing in the BL and WL. The memory cell did not have an access device as only the device at the intersection of the BL and WL experiences enough magnetic field to induce switching, with the unselected devices on either lines see only half the required field (a concept referred to as 'half-select'). The read operation in ferrite core memories were also destructive, and the write-after-read concept employed in FeRAM was initially developed for ferrite core memory. The ferrite cores that were built in the final generations were about $0.2 - 0.3$ mm in diameter and the chips achieved a cycle time in the range of $100 - 200$ ns. This was the most commonly used computer memory device in the period ranging from 1950s-1970s [152].

During the next 20 years, many ideas were pursued for improving the performance of magnetic memories. Notable among them are the use of magnetic bubbles or domains for data storage (bubble memory [62]), use of magnetostriction for moving domain walls (twister memory [20]) and electroplating nickel-iron permalloy on wires (plated wire memory [43]). However, none of them found wide acceptance as their cost and performance was eclipsed by those of DRAM chips.

The discovery of giant magneto-resistance (GMR) and the introduction of magnetic spin-valve devices in the 1990s lead to a resurgence in interest in MRAMs because of their potential for deployment as a universal memory [148, 198]. The simplest device structure comprises of a thin non-magnetic material sandwiched between two ferro-magnetic layers; the earliest demonstration was in a trilayer stack of Fe/Cr/Fe [11]. The resistance of the stack is found to be larger when the orientations of the ferromagnetic layers are anti-parallel compared to the parallel configuration (Figure 1.6 A). If the middle layer is made of an electrical insulator, the overall impedance as well as the resistance contrast for the two configurations of the ferromagnetic layers increases further – the conductivity modulation is due to a quantum mechanical phenomenon called tunnelling magneto-resistance (TMR) [124, 142]. Typical insulators used in such magnetic tunnel junctions (MTJ) include Al_2O_3, GeO and MgO and commonly used ferro-magnetic layers are $Co_{90}Fe_{10}$, $Co_{40}Fe_{20}B_{20}$ and $Ni_{80}Fe_{20}$.

GMR is based on the coupling and scattering of spin polarized electrons in the magnetic structure while TMR relies on spin - polarized electron tunnelling between the two ferromagnetic layers through the insulator. GMR (defined as $(R_{AP} - R_P)/R_P$), where R_{AP} and R_P are the resistances for anti-parallel (AP) and parallel (P) alignment of the magnetic layers) typically lie in the range of $20 - 30\%$ while TMR values up to 600% has been demonstrated in MgO based systems [63].

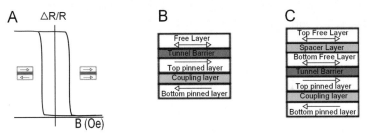

Figure 1.6:   (A) The resistance of magnetic tunnel junctions (MTJ) depends on the relative orientation of the ferro-magnetic layers, up to 600% difference between the two configurations have been demonstrated. Also shown are the simplified schematic of (B) the conventional MRAM stack and (C) the toggle MRAM stack. (Adapted from [18, 210]).

The earliest proposals for building magnetic memory arrays based on these new physical phenomena called for the integration of MTJs in cross-bar arrays, either in series with a select diode or by itself. Because of the low on-off ratio of the memory element, it is better to have an access device in series with the MTJ, especially for configuring large memory arrays [46]. For building memory cells, a multi-layered stack of magnetic materials is engineered such that the magnetic polarization of one of the ferromagnetic layers is held pinned, while the direction of polarization of the other layer could be flipped between two directions (Stoner-Wohlfarth switching) (Figure 1.6 B). In a more advanced version called the toggle MRAM (Figure 1.6 C), instead of a single free layer, two magnetic layers separated by a non-magnetic spacer layer is used – this configuration has better immunity to half-select disturb and thermal fluctuations [210]. The magnetic layers are typically $2 - 5$ nm thick while the insulator layer is $1 - 2$ nm thick. Note that this is a highly simplified, conceptual picture of the MTJ stack; it is typical to have $8 - 10$ distinct layers of ferromagnetic, anti-ferromagnetic, dielectric and other interfacial layers to ensure good thermal stability.

The memory cell size in early demonstrations of conventional MRAMs was typically in the range of $40 - 50$ F^2, with a separate write word-line (WWL) in addition to the conventional BL and WL [18] (Figure 1.7 A). For programming, the select transistor is turned off, and the orientation of the free layer is flipped based on the magnetic fields set up by the programming currents flowing through the BL and WWL. For reading, the WL is turned on and a current proportional to the MTJ conductance is sensed at the BL. Both programming and read times under 10 ns, as well as write endurance in excess of 10^{14} cycles could be achieved in these demonstrations [39, 40].

The main limitation with Stoner-Wohlfarth and toggle MRAMs is that the required programming current density increases beyond 10 MA/cm^2 once the lateral dimension of the magnetic layers scale below 100 nm [150]. One of the alternatives that was pursued to get around this challenge was to temporarily lower the writing field of the MTJ by Joule heating; this is done by applying current pulses through the memory element and surrounding metal wires to locally heat the MTJ volume [15].

Attention in the last 10 years have shifted to a novel form of MTJ programming, utilizing the spin-torque transfer (STT) effect which was theoretically predicted in 1996 [17, 190]. The basic idea behind this scheme is that spin polarized currents caused by the interaction (transmission or reflection) with the pinned ferro-magnetic layer can exert a spin-torque on the free ferro-magnetic layer and flip its orientation. This scheme is inherently superior as the mechanism of switching is local (hence does not need an extra write WL) and the required programming current scales proportionately with physical scaling.

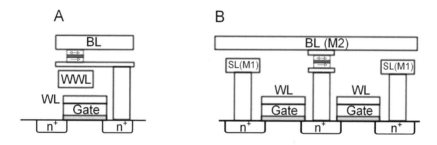

Figure 1.7: (A) Conventional MRAM cell comprises of an FET to prevent sneak paths during reading, and a separate Write WL (WWL) to program the memory bit. (Adapted from [18]). (B) An STT-RAM cell does not require the write-WL as it is programmed and sensed by currents flowing through the MTJ. Most STT-RAM chip demonstrations so far utilizes two FETs connected in parallel with the MTJ, though this is not a requirement (Adapted from [199]).

Early demonstrations of STT-RAM have been promising, with cell sizes down to 14 F^2, programming pulse widths less than 10 ns, excellent thermal stability and cell endurance of about 10^{12} cycles [57]. Typical memory cells consist of two transistors connected in parallel with the MTJ to provide the current necessary for programming (Figure 1.7 B). Bi-directional currents are required for programming – to obtain the parallel alignment of the magnetic layers, programming current flows from the free layer to the pinned layer and in the opposite direction to obtain the anti-parallel alignment [33, 199]. The required programming current density is in the range of $1 - 5$ MA/cm^2, with the potential for further reduction with the use of perpendicular magnetic materials instead of the usual in-plane materials [211]. At the time of writing, the largest STT-RAM chip demonstrated in literature was a 64 Mbit chip in 54 nm technology, employing a 2-transistor, 1 MTJ cell; programming and read speeds less than 20 ns were achieved [33].

Going forward, STT-RAM has the potential to grow into a universal memory because of its fast programming and read capability. However, careful engineering efforts will be required to decrease the effective cell size and also maintain the uniformity and reliability of multi-layered MTJ stacks over the entire wafer. Improving the on-off ratio of the MTJs will also boost the potential for MLC capability in STT-RAM [67].

1.3.4   RESISTIVE RAM

The earliest reports of memory switching in what would today be classified as a Resistive RAM or RRAM date back to the early 1960s [48, 96]. Stable resistive states with an on-off ratio exceeding 10^5 was observed in the oxides of Ni and Si; it was proposed that filament growth or re-distribution of impurity species within the active volume could explain the switching phenomena. Since then, a wide variety of materials have been shown to exhibit conductivity modulation and memory effects under high electric field conditions (typically $> 10^6$ V/cm). However, there is no unified theory to explain the fundamental basis of memory switching – depending on the material system, a wide variety of mechanisms have been proposed to explain the observed switching characteristics [70].

Before we discuss the various categories of resistive memory candidates, it is instructive to identify some characteristic features that are common to these devices. The memory device consists of an insulating material that is sandwiched between two metal electrodes; the typical insulating materials used are binary oxides of metals such as Ni, Ti, Zr, Cu, etc. [49, 189, 194, 205], transition metal oxides such as $Pr_{0.7}Ca_{0.3}MnO_3$, $SrZrO_3$, $SrTiO_3$ [75], or solid state electrolytes such as Ge_xSe_{1-x}, Ag_2S, Cu_2S [91]. The choice of the electrode material (e.g., Pt, Ti, Ag, Cu) is also important as the metal work function plays a crucial role in the memory switching effect [208]. Most of these memory candidates exhibit bi-polar switching, i.e., the switching behaviour depends on the polarity of applied electric field or current, primarily because of some spatial asymmetry in the device or material structure [104]. However, uni-polar operation is not uncommon either [56, 114]. Most devices in the RRAM family require a forming process, where a substantially different programming condition is first applied to an as-fabricated device, before it exhibits memory switching effects [215]. Also, for many device candidates, the magnitude of the on-state resistance is determined by the compliance limit of the programming current [47], but there are exceptions to this as well [113]. Many RRAM devices also exhibit a high degree of non-linearity in its I-V characteristics, in both the off and on states; this has buoyed the hopes for a true cross-bar array implementation that requires no external access devices [30].

We can classify the various RRAM devices based on some commonly proposed switching mechanisms.

- Redox reaction
  In this class of devices, memory switching is based on reduction/oxidation (redox) electrochemical reactions that result in formation and dissolution of conducting filaments (e.g., NiO [82]) or a more homogeneous modification of conductivity at the interface between the active volume of the material and the metal electrode (e.g., Sm/$La_{0.7}Ca_{0.3}MnO_3$) [54, 173]. Redox reaction

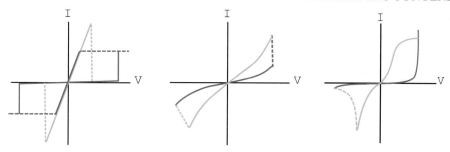

Figure 1.8: Typical I-V curves observed for RRAM devices. (Adapted from [173, 213]).

in these devices can be either cation mediated (e.g., $Cu/Ta_2O_5/Pt$ [200]) or anion mediated (e.g., $SrTiO_3$ [13]), depending on the electrodes and the active material. The memory switching seen in $Pt/TiO_2/Pt$ based memristor devices also falls in this category, as the conductivity modulation is explained on the basis of field induced movement of the boundary between the conductive TiO_{2-x} and the non-conductive TiO_2 within the active volume [207, 213]. The class of RRAM devices called programmable metallization cell (PMC) based on the growth and dissolution of a silver or copper electrodeposit in solid state electrolytes could also be considered to belong in this category [91].

- Charge injection and trapping
  It is postulated that the memory effect seen in $Ti/Pr_{0.7}Ca_{0.3}MnO_3$ [172], Ti/La_2CuO_4 [174] and ZnO [103] based RRAMs could be attributed to the reversible voltage induced modulation of the Schottky-like barrier height at the metal/oxide interface due to charge injection and trapping.

- Mott metal-insulator transition
  Band theory of solids is essentially derived assuming periodic lattice structure of atoms in crystals, neglecting the electrostatic interaction between electrons. These electron-electron interactions are important in explaining the poor conductivity of transition metal oxides, in spite of having partially filled electron bands. Conductivity of such materials could be modulated by charge injection or doping – this transition from a strongly electron correlated insulator into a weakly electron correlated metal is called Mott metal-insulator transition [65]. It is believed that this mechanism is behind the resistance switching effects seen in $Au/Pr_{0.7}Ca_{0.3}MnO_3/Pt$ based RRAM devices [83].

Research efforts in the field of RRAMs are beginning to mature from basic materials physics and device engineering to large array scale demonstrations. The largest chip demonstration among this at the time of writing is a 64 Mbit 130 nm RRAM chip fabricated by Unity Semiconductors [30]. The chip consisted of 4 stacked layers of memory cells based on conductive metal oxide technology (cell size $0.168\mu m^2$, no access devices) and could achieve programming and erase based on half-select biasing schemes with pulse widths in the $1 - 10\mu s$ range. There are also small 4-Mb test macro demonstrations based on CuTe-based conductive bridge RAM [7] and HfO_x based bipolar RRAM [105]. These demonstrations have achieved sub-50ns read and programming time, 2-bits per cell MLC capability and good endurance characteristics ($\sim 10^8$ cycles) [138, 186].

As is evident from the discussion above, there are a plethora of material/device systems that could be potential RRAM based next generation memory candidates, and so far, none has emerged as a clear winner. There is considerable debate on the fundamental mechanism of memory switching in these devices as well as its scalability. The forming process necessary for many of these candidates is also an added burden in their paths to commercialization. The success of RRAM will depend on establishing a front runner among these candidates and unambiguously demonstrating its scalability and endurance potential.

1.3.5   EMERGING MEMORY CANDIDATES AT A GLANCE

We now list the key attributes and features of the emerging memory candidates discussed so far, as well as that of PCM in Table 1.2. This comparison shows that PCM has demonstrated significant progress and maturity to becoming a viable next generation memory candidate. In addition to large chip-scale demonstrations (1 Gb) in advanced technology nodes (down to 45 nm), there is significant understanding about the basic material and device physics, scaling theory and reliability of phase change memory, which will be the topic of our discussion in the following sections.

Table 1.2: Comparison of metrics of largest chip demonstrated for various memory technologies. Note that these may not be representative of the best-in-class demonstrations in other metrics.

Metric	Flash	FeRAM	STT-RAM	RRAM	PCM
Technology node	27 nm	130 nm	54 nm	130 nm	45 nm
Half-pitch, F	27 nm	225 nm	54 nm	200 nm	52 nm
Memory area	$0.00375\mu m^2$	$0.252\mu m^2$	$0.041\mu m^2$	$0.168\mu m^2$	$0.015\mu m^2$
Cell size	$5.1F^2/3$	$5F^2$	$14F^2$	$4.2F^2/4$	$5.5F^2$
Chip size	64 Gb	128 Mb	64 Mb	64 Mb	1 Gb
Write speed	7 MBps	83 ns	< 15 ns	$1 - 10$ ms	$100 - 500$ ns
Read speed	200Mbps	43 ns	< 20 ns	100 ms	85 ns
Vcc	$2.7 - 3.6$ V	1.9 V	1.8 V	$3 - 4$ V	1.8 V
Source	[102]	[187]	[33]	[30]	[181]

1.4   PHASE CHANGE MEMORY

The origins of phase change memory can be traced back to the pioneering work of S. Ovshinsky on the physics of amorphous and disordered solids in the late 1960s; he proposed that the reversible transitions between the ordered and disordered phases of chalcogenide glasses (alloys consisting of elements from the chalcogen family, e.g., O, S, Se, Te, etc.) that exhibit markedly different resistivity could be utilized to form solid state memory devices [140]. The transitions between the two phases of these novel materials could be achieved either electrically or via optical laser pulses [139]. This was almost immediately followed by what could be considered as the first array demonstration of PCM – a 256 bit "read mostly memory" configured as a 16×16 array of these 'ovonic devices' in series with integrated silicon diodes was demonstrated in 1970 [130, 131]. Programming was achieved by electrical pulses, with the key insight that the energy-time profile (and not just the energy) of the applied pulse is critical in determining the final state of the memory device. The initial assessments of the technology were very optimistic; it was projected that "cell densities approaching one million bit per square inch appear possible" [147].

In spite of this and other similar demonstrations [184], PCM failed to become the accepted choice for a solid state memory as the required programming energy was much larger than that for DRAM and flash memory. However, the good news is that the energy required for programming scales in proportion with the active volume of the phase change element, and is projected to scale below the pico-Joule range once the diameter of the active volume shrinks below 50 nm [94]. This, along with the difficulties for scaling flash memories has now opened a window of opportunity for PCM to enter the marketplace.

1.4.1   PCM MATERIALS/DEVICE PHYSICS

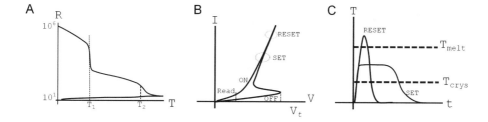

Figure 1.9: (A) Typical R-T (normalized) curve of chalcogenide films show that the resistivity of amorphous phase is $5 - 6$ orders of magnitude higher than the polycrystalline phases. T_1 and T_2 are the temperatures where the transition to f.c.c and h.c.p phases take place. (B) I-V curves observed for PCM devices show that in the on state, the device behaves like a non-linear resistor. In the off state, the devices undergoes threshold switching at a critical bias (V_t). (C) Ideal thermal profiles generated within the cell during SET and RESET programming. (Adapted from [26, 209]).

A wide variety of materials exist in ordered (poly-crystalline or crystalline) and disordered (amorphous) form; however, chalcogenide phase change materials (e.g., $Ge_2Sb_2Te_5$, GST in short) are ideally suited for realizing the next generation memory technology as it satisfies the following properties.

- Resistivity contrast between poly-crystalline and amorphous phase
  The as-deposited (at room temperature) phase of the alloy is amorphous, with typical sheet resistivity greater than $10^7 \, \Omega/\text{sq}$. On annealing, the resistivity of the film decreases as a function of temperature, until a drastic drop in resistivity is observed at about 150 °C and 350 °C. At 150 °C, the resistivity drops by a factor of ~ 100, while at 350 °C, the drop is about a factor of 10 (Figure 1.9A). This drastic change in resistivity upon annealing has been attributed to the structural changes in the crystal ordering, as confirmed by X-ray diffraction analysis – the material first crystallizes to a metastable f.c.c lattice and then to a stable h.c.p phase at these two temperatures, respectively [44]. Even though the resistance contrast between the ordered and disordered phases in thin films can be in the range of $10^6 - 10^8$, typical on-off ratio that is achieved in phase change memory devices is in the range of $100 - 500$ [209].

- Stable phases at operating temperatures
  The amorphous and the h.c.p phases of the alloy are thermodynamically stable and memory cells built based on these materials can retain their state for extended periods of time (> 10 years), even at elevated temperatures (~ 85 °C). It has also been demonstrated that as the physical dimensions of the materials scale, it is possible to engineer the alloy composition or interfaces such that these characteristic features necessary for non-volatile data storage mechanism are maintained [71, 166].

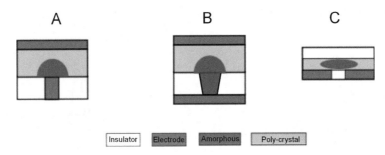

Figure 1.10: In order to maintain reasonable programming currents, phase change memory devices are configured such that one of the critical dimensions of the current flow path is in the range of $10 - 50$ nm. In the mushroom cell (A), a thin film of chalcogenide material is contacted by electrodes (typically TiN), with the diameter of the bottom electrode contact (BEC) in the range of $10 - 50$ nm. In the pore cell (B), a nanoscale pore (dimension $10 - 50$ nm) is filled with the chalcogenide material. In the bridge cell (C), a thin layer ($5 - 50$ nm thick) of chalcogenide material is deposited on top of patterned electrodes that are few 10s of nm apart. (Adapted from [26, 209]).

- Fast programming
  The phase change memory cell is typically configured as a thin film of the phase change material that is contacted on either sides by metal electrodes – some typical device structures

are shown in Figure 1.10. The transition between the two states in such a compact, nano-scale memory cell could be induced in the time scale of $5-500$ ns by electrical pulses that cause Joule heating within the active volume of the material. To program the memory cell to a high resistance state (RESET operation), an electrical pulse is applied to raise the temperature of a substantial portion of the active volume of the chalcogenide alloy beyond its melting point (660 °C) (Figure 1.9C). The current density required for this is in the range of $5-20$ MA/cm^2 [51]. Since the thermal time constant of typical memory cells built in the sub-90 nm technology generations is less than 10 ns, thermal steady state conditions can be attained in similar time scales and the molten volume can be quenched to the high resistive amorphous state if the fall-time of the RESET pulse is within a few nanoseconds.

Programming the cell back to the low resistance state (SET operation) is also based on Joule heating to re-crystallize the amorphous material back to the poly-crystalline phase, and this typically requires raising the temperature to about 350 °C (Figure 1.9C). However, the bias conditions necessary to create enough current flow and Joule heating to cause this temperature rise would have been enormous, but for a phenomena unique to chalcogenide phase change materials called threshold switching (originally also called ovonic threshold switching or OTS) [167]. Though the underlying physical mechanism is not clearly established, it is observed that the electrical conductivity of the amorphous material drastically and momentarily increases beyond a certain critical electric field ($\sim$ 50 V/μm for GST) due to an electronic (and not atomic) transition [3, 92]. The bias voltage where this transition occurs is denoted as V_t, typically about $0.8-1.2$ V (Figure 1.9B). Thus, to SET a cell, an electric pulse is applied such that initially the bias across the phase change volume exceeds V_t making the amorphous volume conduct. The current flow through the device is then controlled (for instance by suitably biasing the access device) so as to heat the amorphous volume to the crystallization temperature and anneal it back (Figure 1.9C). It is found that the one of the best methods to get a tight distribution of SET resistances in a collection of cells is to apply a pulse with a slow ramp down time so that the cell temperature is reduced gradually (in $50-500$ ns). This rate of ramp down is primarily determined by the speed of crystallization of the phase change material, which can be either dominated by nucleation of crystal seeds or by growth of crystal front [217]. Commonly used chalcogenide Ge$_2$Sb$_2$Te$_5$ is a nucleation dominated material [151], while Te free GeSb alloys are growth dominated [202] and can be doped to combine fast crystallization speed with high crystallization temperature, buoying the hopes for a fast ($<$ 100 ns programming) and non-volatile phase change memory technology [29].

- Large endurance
The chalcogenide material can be cycled between the amorphous and poly-crystalline states a large number of times. Although some atomic diffusion and segregation is expected at high temperatures which could potentially alter the alloy composition or in worst case create shorts and voids within the device structure [161, 171], these deleterious effects can be minimized to extend the lifetime of the memory cells to 10^6-10^8 program/erase cycles. There is active

research in the materials and device engineering community to improve this further by designing programming schemes and alloy composition of the phase change material, specifically because there is some evidence that endurance loss mechanisms are dependent on the time the material spends in the molten state [52].

Also note that the high resistive amorphous phase has poor reflectivity compared to the low resistive poly-crystalline phase of the material in the $400 - 800$ nm region of the electro-magnetic spectrum – this contrast in optical reflectivity is utilized to make CD-RW and DVDs [137].

1.4.2   PHYSICS OF PCM PROGRAMMING AND SCALING THEORY

The physics of programming of PCM devices is governed by the heat diffusion equation, which states that the rate of change in temperature, $T(r, t)$ is determined by the balance between energy input by Joule heating and thermal diffusion losses.

$$dC_p \frac{dT(r,t)}{dt} = \rho |J(r,t)|^2 - \nabla(-\kappa \nabla T(r,t)) \qquad (1.1)$$

where ρ, κ, d and C_p are the electrical resistivity, thermal conductivity, density and specific heat capacity of the material. $J(r, t)$ denotes the current density at spatial point r at time t. It is important to note that these material parameters are themselves functions of temperature, crystalline fraction, etc., and non-linear finite element difference methods are necessary to actually solve for the heating profile within the structure.

Table 1.3: Scaling trends for phase change memory assuming isometric scaling and neglecting non-linear effects.

Metric	Scaling trend
Physical dimension	$1/\alpha$
Current density	α
Thermal time constant	$1/\alpha^2$
Programming current	$1/\alpha$
Electric field	α
Voltage drop across cell	1
Power	$1/\alpha$
Device resistances	α
Melt radius	$1/\alpha$
Threshold voltage	$1/\alpha$

However, analytical models built based on simplifying assumptions can provide crucial inputs for memory cell operation and design. For instance, it has been shown that there is an approximate linear relationship between the programming current magnitude and the radius of the molten volume in the mushroom phase change element at steady state, suggesting that the programming current should scale in a linear fashion with device scaling (Figure 1.9A) [160]. The complete scaling theory can be worked out based on equation 1.1 neglecting the non-linear variations in material parameters with temperature [88], a summary of which is provided in Table 1.3. Note that even though the temperature rise needed for programming does not scale with device scaling, the volume of material that is being heated decreases and hence the programming power decreases as well. The electric field and device resistances also increases proportionately with physical scaling, which has implications on device read bandwidth and reliability.

The above analysis assumes that the phase change characteristics of chalcogenide alloys are not size dependent. This is not the case in reality, as material parameters such as resistivity, crystallization temperature and speed, melting point, etc. depends on interface layers and volume-to-surface ratio of the material. For instance, the crystallization temperature is found to increase as the film thickness decreases [165] while the crystallization speed is expected to decrease for nucleation dominated alloys and increase for growth dominated alloys [164].

In this context, it is also worth mentioning that there is considerable interest in the properties of nanowires and nanoparticles made of phase change materials due to their potential to further increase device density and reliability. It has been shown that phase change nano-particles with diameters as small as 20 nm can still undergo phase transition on annealing [166] and that the resistance states obtained in $Ge_2Sb_2Te_5$ nanowires are more stable than that in thin films [107, 122]. PCM bridge devices (Figure 1.10C) built using ultra-thin (up to 3 nm thick) phase change materials have also been shown to exhibit excellent data retention and cycling characteristics [29]. These experiments have shown that phase change materials have excellent intrinsic scaling properties and that devices built of these materials are expected to scale beyond the 22 nm node.

1.4.3 MEMORY CELL AND ARRAY DESIGN

One of the primary objectives of the phase change element (PCE) design is to minimize the programming current while maintaining the device resistances at reasonable values so as to achieve good read and write speeds. The RESET programming operation is the current hungry operation, while the SET programming operation is the time consuming one. Since the RESET current density required for programming is in the range of $5 - 20$ MA/cm^2, at least one physical dimension of typical PCEs are patterned to be in the nano-scale range [9, 64]. It is critical that the variability associated with this feature be minimized as it is intimately related to programming current and device resistances – there are processing techniques that can achieve regular array of nanoscale features that do not inherit the variability of conventional lithography [22, 159]. In addition, the thickness of the chalcogenide as well as surrounding dielectric films are critical in determining the thermal diffusion losses which affects the cooling rate within the active volume as well as thermal cross-talk between adjacent cells [108].

The phase change memory array is configured in a hierarchical manner, with the fundamental block in the array built with up to $1024 - 4096$ bit-lines and word lines. At the junction of each

BL and WL is typically a PCE connected in series with an access device. The most preferred choice for an access device is the CMOS FET [74], as only minor process changes are necessary to modify the characteristics of FETs available in common foundry offerings to suit the needs of a memory array. However, it is often the case that the current drive capability of minimum sized FETs is much lesser than that required for PCM programming, necessitating the use of wider devices, leading to large cell area. Hence, memory designers employ BJTs and diodes as access devices since they have higher current drive per unit area compared to FETs [109]. The diode or BJT based memory arrays also have the added advantage that the BL/WL contact in each line could be shared among $4 - 8$ devices, allowing a further reduction in cell size.

Employing diodes and BJTs as access device, PCM arrays with cell sizes in the range of $5.5F^2 - 11.5F^2$ have already been demonstrated [136, 181]. However, this comes at the cost of added process complexity, primarily because these devices needed to be integrated in crystalline silicon substrates (before the metal levels and memory devices themselves are fabricated). For instance, the diodes in [136] are fabricated on crystalline silicon obtained by selective epitaxial growth and the memory array integration requires 4 extra mask levels in addition to the standard CMOS process flow. The configuration of the array and peripheral circuits also plays a critical role in determining the overall performance of the chip – for instance, higher array efficiency (which means longer BLs and WLs) comes at the cost of array access time [73, 106].

1.4.4   MLC PROGRAMMING IN PCM

Another key aspect that is unique to PCM devices is that it is possible to program the cell to arbitrary intermediate resistance values between the minimum SET resistance ($10 - 50$ KΩ) and the maximum RESET resistance ($> 1 - 5$ MΩ). By varying the amplitude, pulse width or fall time of the electrical pulse, it is possible to control the size of the amorphous volume within the cell, thereby tuning the effective cell resistance [60]. It has been demonstrated that by applying rectangular pulses with increasing amplitude, the cell resistance can be increased proportionately and the cell resistance can be gradually decreased by applying trapezoidal pulses with increasing fall time [112, 129]. These strategies have been employed to demonstrate impressive MLC capability (up to 4 bits/cell) in PCM devices [132].

Iterative programming techniques are necessary to attain tight resistance distributions in MLC PCM [14]. With suitable write-read-verify programming schemes, it is possible to achieve well controlled distributions even for 4-bit MLC demonstrations with a small number ($5 - 6$) of iterations in conventional PCM devices [132]. An alternative strategy for MLC realization is to engineer the cell structure specifically for programming to intermediate resistance levels – examples of this include incorporating multiple chalcogenide layers with varying electrical resistivity [163] or cell structures with distinct parallel current paths [135].

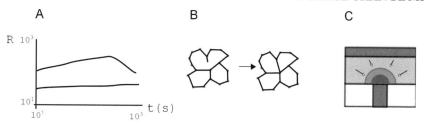

Figure 1.11: The programmed resistance states in PCM devices (A, units in KOhm) drifts upwards initially, and then decreases to-wards the crystalline state resistance. The rate of drift is higher for larger resistances. The programmed resistance increases due thermally activated atomic rearrangements in the amorphous volume (known as structural relaxation, B) and it decreases due to the reduction in the size of the amorphous volume due to crystallization (C). (Adapted from [61, 168, 188]).

1.4.5   RELIABILITY

The real challenge in the commercial deployment of a novel memory technology is in ensuring uniform performance from each of the billions of bits in the array. For this, it is essential to have fundamental understanding related to the failure mechanisms of the technology. Thanks to research efforts in the past five years, we are now in a position to not only predict and characterize various failure mechanisms, but also engineer solutions to circumvent some of these problems for PCM [31, 86, 146].

One of the foremost reliability considerations of any non-volatile memory is data retention, and PCM suffers from two different mechanisms of data loss (Figure 1.11). The programmed resistance level can increase with time due to thermally activated atomic rearrangements in the amorphous volume (known as structural relaxation [61], Figure 1.11B) or can decrease with time due to decrease in size of the amorphous volume due to crystallization [188] (Figure 1.11C). These are not much of a concern for single level PCM as it is for MLC PCM – typical SLC PCM achieves the retention target of 10 years at 85 °C [181]. The upward drift of resistance with time due to structural relaxation is found to obey a power-law model

$$R(t) = R(t_0) \left(\frac{t}{t_0}\right)^{\gamma} \tag{1.2}$$

where γ is a drift-exponent that depends on the initial programmed resistance. Typical value of this drift parameter varies in the range of $0.01 - 0.1$, increasing proportionately with the initial resistance magnitude [168]. Strategies to minimize and manage data retention loss due to drift is a topic of active research and there are many proposals to address this issue [10, 212]. One strategy to minimize the data loss due to crystallization has been to engineer the chalcogenide alloy composition to increase its crystallization temperature, without deleterious effects on its resistivity and other reliability metrics [93, 166].

Two types of endurance failure mechanisms are observed in phase change memory cells - 'stuck-set' and 'stuck-reset' cells [86]. It is believed repeated cycling can cause atomic segregation, resulting in Ge depletion and Sb accumulation, resulting in the cell being stuck at a low resistance state [161]. Stuck-reset failure occurs when the chalcogenide alloy gets physically separated from the electrode, creating a void in the critical current path [196]. It is also observed that the on-off ratio decreases with cycling, and the rate of this degradation depends on the time spent by the active volume in the molten state [52]. With careful materials processing and device engineering [216], it is possible to delay the onset of such degradation mechanisms and attain in excess of 10^8 program/erase cycles [181].

Program disturb and read disturb are two other issues that plague dense non-volatile memories. Since PCM programming involves momentarily raising the cell temperature past $650\,^{\circ}\mathrm{C}$, it is essential to carefully chose the inter-cell dielectric material. It is also important to consider the thermal boundary resistances, especially at nanoscale dimensions where non-linear electron-phonon scattering mechanisms will play a key role in determining the heat transfer profile between neighbouring cells [169]. Electro-thermal simulations indicate that programming thermal disturb is not expected to be an issue even in minimum-pitch PCM arrays down to the 16 nm node [108]. Read disturb in PCM can be minimized by carefully choosing the bias conditions for sensing. Since the typical read voltages are in the range of $0.3 - 0.5$V, there is negligible temperature rise during sensing, and hence does not affect the low resistance state of the cell. If the applied voltage is well below the threshold-voltage of the device, statistical models and experiments indicate that the probability of a memory cell programmed to the high-resistance suffering an accidental threshold switching event can minimized well below 1 ppb [99].

Another reliability concern unique to PCM is random fluctuations in measured resistance levels, which is believed to be caused by the statistical fluctuation in the number and energy density of trap states in the amorphous state of the chalcogenide alloy [42]. Recent studies have shown that a raw bit error rate of $\sim 10^{-4}$ is achievable for 3-bit/cell 1 Mb PCM – there are many efficient ECC strategies that are being pursued to boost the chip level reliability further [34, 98]. It should also be noted that unlike DRAM or SRAM, phase change memory technology does not suffer from soft-errors induced by charge-based radiation effects, making it suitable even for space applications [119].

1.4.6   PCM TECHNOLOGY MATURITY

Even though the idea of chalcogenide material based data storage dates back to the early 1960s, the first publication that revived interest in the development of PCM technology appeared in 2001 [94]. Thanks to the concerted research in the next 10 years in various industrial and academic research laboratories, PCM technology development has progressed rapidly, with large array prototypes (512 Mb - 1 Gb) built in advanced technology nodes (65 nm or 45 nm) now available for initial product sampling. It has also been reported that Samsung and Micron have announced the introduction of PCM memory chips in commercially available products such smartphones and mobile handsets as well as for use in embedded systems and solid state storage subsystems. There has also been development of modeling and design tools for memory circuit design based on PCM [206], not to mention a wide variety of systems and architectural solutions proposed to manage the reliability challenges of PCM.

1.4.7 CONCLUDING REMARKS

From the discussion above, it should be clear that phase change memory has emerged as a leading contender to take the role of the next generation memory technology. Impressive progress has been made in basic materials, device engineering and chip level demonstrations including the potential for MLC programming and 3d-stacking [78]. In addition, there is a large body of knowledge on the failure mechanisms of this technology with active research efforts aimed at suppressing or managing these issues. Unlike other emerging memories, PCM satisfies the many stringent requirements that are necessary for enterprise applications as it combines the best features of NAND flash memory (such as retention) with those of DRAM and SRAM (read and write bandwidths). Supported by architectural and systems level solutions, PCM is poised to enable a revolution in next-generation computational systems, which is the topic of discussion of the following chapters.

CHAPTER 2

Architecting PCM for Main Memories

2.1 INTRODUCTION

The increasing core-counts in modern high performance systems requires that the memory system must scale in capacity in order to accommodate the working set of the programs on all the processors. For several decades, DRAM has been the building block of the main memories of computer systems. However, with the increasing size of the memory system, a significant portion of the total system power and the total system cost is spent in the memory system. Furthermore, scaling DRAM to small feature sizes is becoming a challenge, which is causing the DRAM technology to shrink at a slower rate than the logic technology. Figure 2.1 compares the trend for on-chip core counts with respect to DRAM DIMM capacity (redrawn from Lim [80]). The processor line shows the projected trend of cores per socket, while the DRAM line shows the projected trend of capacity per socket, given DRAM density growth and DIMM per channel decline.

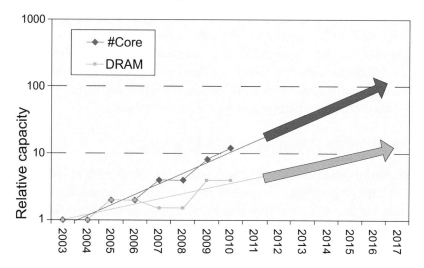

Figure 2.1: Trends leading towards the memory capacity wall. (Figure adapted from [80].)

If the trends continue, the growing imbalance between supply and demand may lead to memory capacity per core dropping by 30% every two years, particularly for commodity solutions. If not addressed, future systems are likely to be performance-limited by inadequate memory capacity. Therefore, architects and system designers are motivated to exploit new memory technologies that can provide more memory capacity than DRAM while still being competitive in terms of performance, cost, and power. Two promising technologies that can bridge the gap between DRAM and Hard disk are *Flash* and *PCM*.

Figure 2.2 shows the typical access latency (in cycles, assuming a 4GHz machine) of different memory technologies, and their relative place in the overall memory hierarchy. A technology denser than DRAM and access latency between DRAM and hard disk can bridge the speed gap between DRAM and hard disk. Flash-based disk caches have already been proposed to bridge this speed gap, and to reduce the power consumed in HDD.

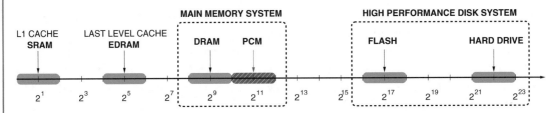

Figure 2.2: Typical access latency of various technologies in the memory hierarchy.

However, with Flash being two to three orders of magnitude slower than DRAM, it is still important to increase DRAM capacity to reduce the accesses to the Flash-based disk cache. The access latency of PCM is much closer to DRAM, and coupled with its density advantage, PCM is an attractive technology to increase memory capacity. Furthermore, PCM cells can sustain 1000x more writes than Flash cells, which makes the lifetime of PCM-based memory system in the range of years as opposed to days for a Flash-based main memory system.

2.2 PCM: BENEFITS AND CHALLENGES

PCM is expected to be a scalable technology with feature size comparable to DRAM cells. Furthermore, a PCM cell can be in different degrees of partial crystallization thereby enabling more than one bit to be stored in each cell. Several prototypes have demonstrated multiple bits per cell. Given the scaling advantage and multiple bits per cell, PCM can enable a larger main memory system in the same area. Furthermore, the non-volatile nature of PCM means significant leakage power savings.

Unfortunately, PCM suffers from three major drawbacks. First, higher read latency. Current projections show that PCM is about 2x-4x slower than DRAM, which means a PCM only system is likely to have much higher memory latency, reducing performance. Second, higher write latency and energy. The latency to write a line in PCM memory is more than an order of magnitude higher than in DRAM. Furthermore, writes consume significantly high energy. Therefore, a heavy write traffic would results in much reduced performance and higher power consumption. And, finally, PCM has limited write endurance. PCM cells are projected to endure 10-100 million writes, which means that a lifetime of PCM system would be limited due to endurance related wear-out. Several recent architecture studies have looked at mitigating these drawbacks of PCM in order to make it viable as a main memory candidate. This chapter describes three such proposals. The content of Section 2.3 is derived from [100], Section 2.4 is derived from[218], and Section 2.5 is derived from[155].

2.3   PCM TAILORED ARRAY ORGANIZATION

Lee et al. [100] looked at array architecture and buffer organizations that are well suited to PCM. Similar to conventional memory organization, PCM cells can be organized into banks, blocks, and sub-blocks. However, one can benefit in terms of power and performance by making PCM-specific optimizations to the memory array architecture. For example, PCM reads are non-destructive, therefore it is not necessary to write back the row buffer if the data in the row buffer has not been modified.

Conventional DRAM architecture implement both sensing and buffering using cross-coupled inverters. PCM architectures can separate sensing and buffering, where sense amplifiers drive banks of explicit latches. Such a decoupling can provide much greater flexibility in row buffer organization by enabling multiple buffered rows. Separate sensing and buffering enables multiplexed sense amplifiers too. Multiplexing also enables buffer widths narrower than the array width, which is defined by the total number of bitlines. Buffer width is a critical design parameter as it determines the required number of expensive current sense amplifiers.

The number of sense amplifiers decreases linearly with buffer width, significantly reducing area as fewer of these large circuits are required. The area thus saved can be used to implement multiple rows with latches much smaller than the removed sense amplifiers. Narrow widths reduce PCM write energy too because each memory write is of a much smaller granularity. However, it can also negatively impact performance especially for applications that have high spatial locality. Fortunately, some of the performance penalty may be mitigated by a higher buffer hit rate because of larger number of row buffers.

Lee et al. [100] compared the delay and energy characteristics of different buffer design space for PCM arrays. They observed that a knee that minimizes both energy and delay was: four 512B-wide buffers, instead of the single 2KB buffer for DRAM. Such an organization reduced the PCM delay and energy disadvantages from 1.6x and 2.2x to 1.1x and 1.0x, respectively, making PCM competitive with respect to DRAM.

2.4   FINE-GRAINED WRITE FILTERING

The write operation in PCM is expensive in terms of energy and can limit the usable lifetime of the system. Therefore, techniques that can reduce the write traffic to PCM can not only reduce energy consumption but also increase system lifetime. In a conventional DRAM access, a write updates the content of an entire row (also called a page) of a memory bank. Every bit in the row is written once.

However, a great portion of these writes are redundant. That is, in most cases, a write into a cell does not change its value. These writes are hence unnecessary, and removing them can greatly reduce the write frequency of the corresponding cells. Figure 2.3 shows the percentages of bit writes that are redundant for different benchmarks [218]. The percentage is calculated as the number of redundant bit-writes over the total number of bits in write accesses. The 'SLC' series represents redundant bit-writes in a single level PCM cell, i.e., each cell stores either '0' or '1'. The 'MLC-2' and 'MLC-4' series represent multi-level PCM cells of 2 and 4-bit width. That is, each cell stores 4 (MLC-2) or 16 (MLC-4) binary values.

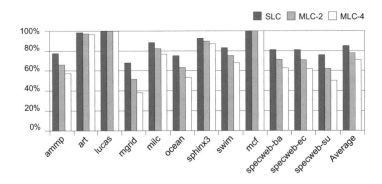

Figure 2.3: Percentage of redundant bit-writes for single-level and multi-level PCM cells. (Figure adapted from [218].)

There are several partial write techniques in literature that have tried to reduce the redundant writes in PCM. For example, writing back only modified words in a line to memory. However, fine grained techniques that can suppress redundant writes at bit level are much more effective than such word level filtering. Zhou et al. [218] proposed light weight circuitry to implement redundant bit removal. Removing the redundant bit-write is implemented by preceding a write with a read. In PCM operations, reads are much faster than writes, so the delay increase here is less than doubling the latency of a write. Also, write operations are typically less critical than read operations, so increasing write latency has less negative impact on the overall system performance. The comparison logic can be simply implemented by adding an XNOR gate on the write path of a cell, as illustrated in Figure 2.4. The XNOR output is connected to a pMOS which can block the write current when the write data equals the currently stored data. The XNOR gate is built based on pass-transistor logic, whose simple structure guarantees both small delay and negligible power dissipation. Given that frequently written bits tends to be spatially correlated, Zhou et al. also proposed the *row shifting* scheme to make the write traffic uniform across the bits in the line.

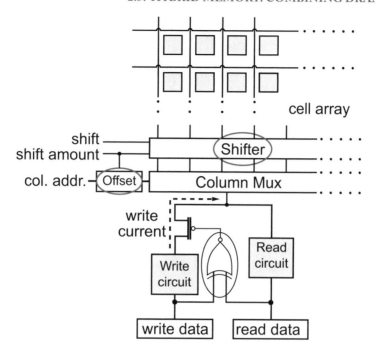

Figure 2.4: Circuitry for Redundant Bit Removal and Row Shifting. (Figure adapted from [218].)

2.5   HYBRID MEMORY: COMBINING DRAM AND PCM

PCM can be used as a replacement of DRAM to increase main memory capacity. However, the relatively higher latency of PCM compared to DRAM can still significantly degrade overall system performance. To address the challenges of PCM, Qureshi et al. [155] proposed a *Hybrid Memory System* consisting of PCM-based main memory coupled with a DRAM buffer. This architecture, shown in Figure 2.5, gets the latency benefits of DRAM and the capacity benefits of PCM. The larger PCM storage can have the capacity to hold most of the pages needed during program execution, thereby reducing disk accesses due to paging. The fast DRAM memory will act as both a buffer for main memory, and as an interface between the PCM main memory and the processor system. Given locality in access streams, a relatively small DRAM buffer (approximately 3% size of the PCM storage) can bridge most of the latency gap between DRAM and PCM.

In a hybrid main memory organization, the PCM storage is managed by the Operating System (OS) using a Page Table, in a manner similar to current DRAM main memory systems. The DRAM buffer is organized similar to a hardware cache that is not visible to the OS, and is managed by the DRAM controller. The buffer has SRAM-based tag-store that contains information about dirty bits and replacement policy. For simplicity, it is assumed that both the DRAM buffer and the PCM

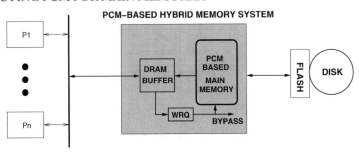

Figure 2.5: PCM-based Hybrid Memory System.

storage are organized at a page granularity. The DRAM cache reduces the read and write traffic to PCM. The write traffic is visible to PCM when dirty pages are evicted from the DRAM buffer. As PCM write latency is high, a write queue (WRQ) is provided to buffer the dirty pages evicted by the DRAM buffer to be written to PCM.

2.5.1   OPTIMIZATIONS FOR HYBRID MEMORY

Qureshi et al. [155] also proposed write filtering techniques that further reduce the impact of writes to the performance, energy, and lifetime of PCM. The first technique *Lazy Write Architecture* avoids the redundant write to dirty pages. It does so by installing all pages in DRAM and writing the pages to PCM only on DRAM eviction. The second technique *Line Level Writeback (LLWB)* exploits the fact that not all lines in a dirty page are dirty. It keeps per-line dirty bits and only writes the dirty lines from DRAM to PCM on eviction. The third scheme *Page Level Bypass*, bypasses pages from PCM for streaming applications, thereby avoiding writes completely for applications that have poor reuse.

2.5.2   PERFORMANCE OF HYBRID MEMORY

Qureshi et al. [155] evaluated the hybrid memory proposal with diverse set of workloads using a full system configuration, assuming that PCM has 4X the density and 4X the latency of DRAM. The benchmark set used consists of database workloads, data mining and numerical kernels, and streaming workloads. Figure 2.6 shows the normalized execution time of four systems: baseline 8GB DRAM, 32GB PCM-Only system, 32GB DRAM, and the hybrid memory system. On average, relative to the baseline 8GB system, the PCM-only 32GB system reduces execution time by 53% (speedup of 2.12X) and the 32GB DRAM only system reduces it by 70% (speedup of 3.3X). Whereas, the hybrid configuration reduces it by 66.6% (speedup of 3X). Thus, the hybrid configuration provides the performance benefit similar to increasing the memory capacity by 4X using DRAM, while incurring only about 13% area overhead while the DRAM-only system requires 4X the area. Qureshi et al. also reported significant savings in energy and power with PCM-based hybrid memory compared to a DRAM-only system further emphasizing that hybrid memory is a practical and power-efficient architecture to increase memory capacity.

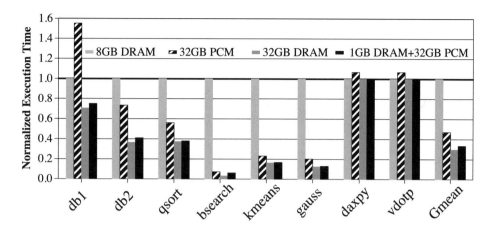

Figure 2.6: Execution Time (normalized to 8GB DRAM).

2.6   CONCLUDING REMARKS

The trend towards increasing core counts in future systems means that the importance of large capacity main memory system is only likely to increase. PCM offers a practical means of scaling memory capacity in a power-efficient manner. In this chapter we discussed initial set of studies that proposed PCM-based memory system, either as a DRAM replacement or as hybrid memory. The non-volatile nature of PCM also enables it to be used for seamless checkpointing and for enabling quick restart in case of failure. Efficiently architecting PCM for main memory system and leveraging its non-volatile property for power-efficiency and reliability continues to be a key research topic. One of the challenges in architecting PCM systems is the slowdown due to contention from long latency writes. The next chapter describes solutions to address that challenge.

CHAPTER 3

Tolerating Slow Writes in PCM

3.1 INTRODUCTION

A characteristic of PCM, similar to most other non volatile memories, is that it has write latency much higher than read latency. A write latency of 4x-8x compared to read latency is not uncommon. A higher write latency can typically be tolerated using buffers and intelligent scheduling if there is sufficient write bandwidth. However, once a write request is scheduled for service from a PCM bank, a later arriving read request for a different line to the same bank must wait until the write request gets completed. Thus, the write requests can increase the effective latency for read requests. Read accesses, unlike write accesses, are latency critical and slowing down the read accesses has significant performance impact. This chapter quantifies the slowdown due to long latency writes and then describes simple extensions to a PCM controller that alleviates most performance lost due to slow writes. The content of this chapter is derived from [157].

3.2 PROBLEM: CONTENTION FROM SLOW WRITES

Qureshi et al. [157] quantified the contention of long latency writes on read requests. Figure 3.1 shows the effective read latency (average) and performance for their baseline PCM system with contention-less read latency of 1000 cycles (the configuration has 32 banks, each with a large write queue, details in [157]). The write latency is assumed to be 8000 cycles and the baseline uses read priority scheduling. For comparison, configurations where the baseline does not use read priority scheduling, writes incur same latency as reads, and writes consume zero cycle latency are also shown.

The time that elapses between the time at which the read request enters a PCM read queue and the time at which it finishes service by a PCM bank is defined as *Effective Read Latency*. The effective read latency for the baseline is 2290 cycles, which is 2.3x compared to the contention-less latency of 1000 cycles. Without read priority scheduling the baseline would have latency of 2788 cycles. If write latency was identical to read latency (1000 cycles) then the effective read latency reduces to 1159 cycles, indicating most of the contention is due to writes and that the write caused delay of reads is a problem only in Asymmetric Write Devices (AWD). If all the contention due to writes is removed then there is a potential to improve system performance by as much as 61% (on average).

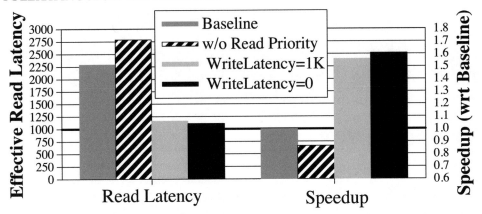

Figure 3.1: Read Latency and Performance Impact of Slow writes. Baseline configuration has read latency of 1K cycles, and write latency of 8K cycles.

3.3   WRITE CANCELLATION FOR PCM

In a conventional memory system, when a read request or a write request is scheduled for service to a bank, that bank is busy until the service is completed. For DRAM systems, the operations for reading or writing are similar in that both types of requests have similar latency and incur a bank write (given that reads are destructive). However, in case of PCM, read and write requests have markedly different latency. When a bank is busy servicing a write request, a later arriving read request waits, which increases the effective read latency. A simple technique to improve read latency in presence of contention from write request is to cancel the write in favor of processing the pending read request [195]. This is called as *Write Cancellation*.

To implement Write Cancellation, PCM devices are equipped with the ability to stop doing the write when the write enable signal is deactivated. The data located at the address being written can end up in a non-deterministic state. However, a copy of the correct data is always held in the write queue until the write completes service. To ensure correctness, a read request that matches a pending write request is serviced by the write queue. Write Cancellation reduces the average effective read latency from 2290 cycles to 1486 cycles.

To avoid starvation of writes, Write Cancellation is performed only when the WRQ is less than 80% full. Even under heavy read traffic the write request will get serviced once the WRQ occupancy exceeds 80%. As soon as the WRQ occupancy falls below 80%, Write Cancellation and read priority are enabled. The writes that are done because the WRQ occupancy exceeds 80% as termed as *forced writes*. Forced writes are done for less than 0.1% of the total writes in the baseline. However, this increased to 2% of total writes when Write Cancellation is employed. Forced writes are detrimental for performance as writes get priority over read requests. Write Cancellation can be controlled to reduce such episodes of forced writes using threshold-based Write Cancellation.

3.4   THRESHOLD-BASED WRITE CANCELLATION

When the service for a write request is close to completion, and a read request arrives, the Write Cancellation policy still cancels the service for that write request. Such episodes of Write Cancellation can be avoided by using a simple time-keeping technique, called *Write Cancellation with Static Threshold (WCST)*. WCST performs Write Cancellation only if the write has finished less than $K\%$ of its service time, where K is a programmable threshold. With K=90%, WCST avoids canceling the write requests that are almost complete, thereby reducing the episodes of forced writes and improving overall performance. To implement WCST, a time-stamp register is added with each WRQ to track the start time of the service for the write request. The difference between current time and the start time is used to estimate if the service is less than K% complete. The total storage overhead to implement WCST for a system with N banks is $(N + 1)$ registers, four bytes each (total overhead of 132 bytes for 32-bank baseline). Note that WCST with K=100% is identical to Write Cancellation. There is significant variation in read latency of different application depending on the value of K. Overall, a value of K=75% achieves an average latency of 1442 cycles.

3.5   ADAPTIVE WRITE CANCELLATION

The best threshold for WCST depends on the WRQ occupancy. If the WRQ associated with a bank has a large number of entries, then Write Cancellation is likely to cause forced writes, so a lower threshold is better as it reduces the likelihood of forced writes. Alternately, if the WRQ is almost empty, a higher threshold can be used as the likelihood of forced writes is low. Based on this insight Qureshi et al. [157] proposed *Write Cancellation with Adaptive Threshold (WCAT)*. WCAT calculates the threshold dynamically based on a simple linear relationship based on the number of entries in the WRQ, using the following equation:

$$Threshold\ (K_\psi) = 100 - \psi \cdot NumEntriesInWRQ \tag{3.1}$$

The parameter ψ is a weight associated with WRQ occupancy. We use $\psi = 4$ which reduces the likelihood of canceling a write request as WRQ occupancy increases, and nearly disables cancellation as WRQ occupancy nears 25 (the WRQ occupancy for forced write is 26). WCAT requires no additional storage overhead compared to WCST, except for minor logic overhead. The average read latency with WCAT is 1400 cycles, and for WCST with best threshold is 1442 cycles. WCAT has reduced latency compared to WCST for all workloads, even when WCST uses the best static threshold (K=75%). This is because WCAT can adapt to different workloads, different phases of the same workload, and has different threshold for each bank.

3.6   OVERHEADS: EXTRANEOUS WRITES

When the service of an on-going write request is canceled, the write request must be scheduled for service again. Thus, the cycles spent in doing the writing for the request just before cancellation are essentially rendered useless. Such episodes of useless write cycles waste energy. WCST and WCAT avoid canceling writes that are almost done, so the average time spent in a useless write is much lower. On average, the number of cycles a bank is servicing write requests increases by 41% with Write Cancellation, by 27% with WCST, and by 22% with WCAT. Thus, WCAT not only improves performance but also reduces the energy overhead associated with Write Cancellation by almost half. The energy overhead of WCAT can be eliminated by a scheme that obviates re-execution of writes that give up their service for processing read requests. Such a scheme is described next.

3.7   PAUSING IN ITERATIVE-WRITE DEVICES

A peculiar characteristic of PCM, as well as other non-volatile memory technologies, is that write time may be non-deterministic, which means two write operations may take different amount of time. This phenomenon is due to write-and-verify techniques, which are necessary to obtain multiple bits per memory cell. The role of a programming technique is to obtain resistance of the PCM element close to that of the desired resistance level. Experimental results show that different cells (and even the same cell at different times) respond differently to the same programming pulse. An interpretation of this behavior in terms of variation and randomness of the crystallization speed has been proposed in [120]. Regardless of the real nature of the uncertainty of PCM response to programming, the acknowledged solution to this problem is the use of a write-and-verify, or *iterative*, programming technique (see, e.g., [132]). A generic write-and-verify algorithm is shown in Figure 3.2.

As the name suggests, a write-and-verify programming technique consists of applying a programming pulse followed by a read stage which is used to either decide whether to stop or to compute the new programming signal to be sent. Programming algorithms which contain an update rule for the programming pulse, such as that in [132], will usually have a small chance of convergence during the first iterations. The probability of convergence then increases as the programming algorithm matches the programming pulses to the actual behavior of the cell. The first few iterations can therefore be seen as a "learning" phase, whose role is that of building an accurate model for the cell which is being programmed, whereas the following iterations are those in which usually the desired resistance level is reached.

Qureshi et al. [157] developed a model to capture iterative writing and reported that a representative 3-bit per cell memory would require on average about 8.3 iterations. This number is compatible with the statistics reported in the literature. For example, Figure 13 of [132] shows the distribution of convergence of one multi-bit PCM cell. As writes are performed at a line-size (256B) granularity, the effective write time for writing one line will be determined by the worst case of all the 684 cells in the line (with 3 bits/cell). Given that about 2% of the cells take 8-9 iterations in [132], the average number of iterations for completely writing a group of 684 cells will be between 8-9.

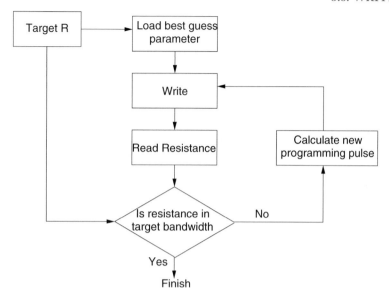

Figure 3.2: Generic algorithm for iterative write.

3.8   WRITE PAUSING

Iterative writing essentially contains steps of writing and checking the resistance value of cells, and rewriting if necessary. The procedure can potentially be stopped at the end of each such iteration for servicing a pending read request for another line in the same bank. Such a scheme is called *Write Pausing*. Write Pausing can significantly improve read latency, especially if the writes contain several iterations, as the end of each iteration represents a potential pause point. At each potential pause point, the device can check if there is a pending read request, service it, and resume the write from the point it was left.

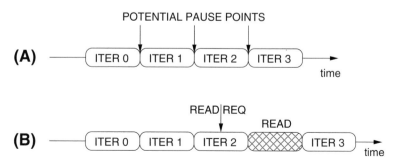

Figure 3.3: Write pausing in PCM: (A) Pause points (B) Servicing reads via Pausing

Figure 3.3(A) shows an example of an iterative write with four iterations. It contains three points at which pausing can be performed. Figure 3.3(B) shows a scenario where a read request arrives during Iteration2 of the write. With Write Pausing, this request can be serviced at the end of Iteration2. Iteration3 of the iterative write begins once the read request completes service. Thus, Write Pausing allows the read to be performed transparently at pause points. Figure 3.4 shows how pausing can be performed within the generic iterative algorithm in Figure 3.2 (the newly added stage is shaded). At each iteration the device recalculates a new programming pulse depending on the state of the PCM cell. However, this writing step can be performed after servicing another read request as well. So, another stage is added to the algorithm that checks if there is a pending read request. If so, the pending read request is performed and the cell writing is resumed as soon as the service for the read request is completed. Note that this argument can be extended to a variety of iterative-write algorithms by identifying suitable pause points.

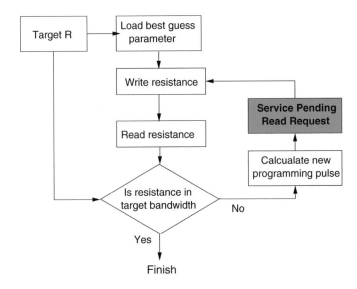

Figure 3.4:   Iterative write algorithm with Write Pausing (newly added stage is shaded).

An ideal pausing scheme can pause the write at any cycle without any loss of programming. Although such a scheme is not practical, it provides an upper bound to compare for Write Pausing, which can pause only at end of each write iteration. Evaluations for Write Pausing show that it reduces the effective read latency by more that half. On average, the baseline has latency of 2365 cycles, Write Pausing has 1467 cycles and Optimal-Pause has 1306 cycles. Thus, Write Pausing gets 85% of the benefit of Optimal-Pause, while incurring minor changes to PCM controller.

3.9   COMBINING WRITE PAUSING AND CANCELLATION

The latency difference between Write Pausing and Optimal-Pause happens because an arriving read must wait until a potential pause point. This latency can be avoided by doing an *Intra-Iteration Write Cancellation (IIWC)*, which can cancel a write iteration, and re-execute that iteration when the write resumes. Figure 3.5 shows an example of the baseline system, Write Pausing, and Write Pausing with IIWC.

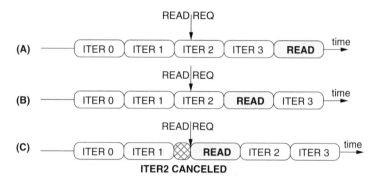

Figure 3.5:   Servicing reads in (A) Baseline (B) Write Pausing (C) Write Pausing + IIWC.

In the baseline system, the incoming read request has to wait for all the write iterations to finish. With Write Pausing, the request waits until the end of Iteration 2, which still incurs latency, albeit small. With Write Pausing + IIWC, the read request starts service almost instantly after it arrives, iteration 2 gets canceled and is re-executed after servicing the read request. IIWC must avoid canceling a write iteration if it is close to completion. So, to implement IIWC, Write Cancellation with Adaptive Threshold is used (with 100% latency being the time to do one iteration). IIWC is less expensive than Write Cancellation as IIWC cancels only a fraction of one write iteration instead of the entire service time of write. On average, Write Pausing has a latency of 1467 cycles, Write Pausing+IIWC has 1323 cycles, and Optimal-Pause has 1306 cycles. Thus, Write Pausing + IIWC obtains almost all of the potential performance improvement, which incurring a negligible write overhead of 3.5% on average.

3.10   IMPACT OF WRITE QUEUE SIZE

The baseline configuration in this study assumed a 32-entry WRQ for each bank. Figure 3.6 shows the relative performance of the baseline and Write Pausing + IIWC as the entries in WRQ is varied from 8 to 256. The performance of the baseline is relatively insensitive to WRQ size, and saturates at 32-entry per WRQ. Write cancellation and pausing, on the other hand, continues to benefit significantly with increase in WRQ size. This is because the baseline services a pending write to completion as soon as the RDQ is empty. Whereas with cancellation or pausing, a pending write may stay in the WRQ for a longer time because of Cancellation and Pausing. While the study used a per-bank WRQ for simplicity, one may get the benefit of a larger per-bank WRQ by using a global WRQ and allocating it to banks based on demand, at the expense of increased complexity.

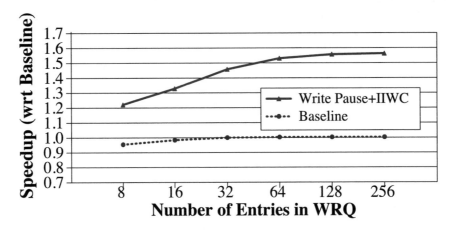

Figure 3.6:   Relative performance of baseline and Write Pause+IIWC as WRQ size is varied.

3.11   CONCLUDING REMARKS

This chapter showed that the contention caused by write requests significantly increases the latency of read requests. It describes Write Cancellation and Write Pausing as means for tolerating slow writes. Other organizational improvements that can reduce the latency or increase bandwidth for write requests can also improve effective read latency. Reducing the write traffic to PCM, for example using delayed write in DRAM cache, can reduce this contention as well. As PCM technology develops, such architecture solutions will be necessary to improve read latency, read bandwidth, write latency, and write bandwidth.

CHAPTER 4

Wear Leveling for Durability

4.1   INTRODUCTION

One of the major challenges in architecting a PCM-based memory system is the limited write endurance, currently projected between $10^7 - 10^8$ [1]. After the endurance limit is reached, the cell may lose its ability to change state, potentially giving data errors. If writes were uniformly distributed to each line in memory, this endurance limit would result in a lifetime of several (4-20) years for a PCM-based memory system [154]. Unfortunately, write accesses in typical programs show significant non-uniformity. Figure 4.1 shows the distribution of write traffic to a 16GB memory system (memory contains 64M lines of 256B each, writes occur after eviction from DRAM cache) for the *db2* workload in a given time quanta. For *db2* most of the writes are concentrated to a few lines. The maximum write count per line is 9175, much higher than the average (64). The heavily written lines will fail much faster and will cause system failure much earlier than the expected lifetime.

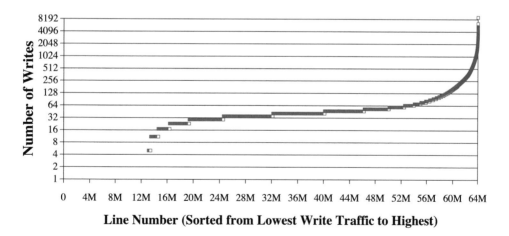

Figure 4.1:   Non-uniformity in write traffic for db2.

Figure 4.2 shows the expected lifetime of the memory system normalized to the case when writes are assumed to be uniform. To avoid the pathological case when only very few lines cause system failure, the system contains 64K spare lines; the system fails when the number of defective lines is greater than the number of spare lines. Even with significant spares, the memory system can achieve an average lifetime of only 5% relative to lifetime achieved if writes were uniformly distributed.

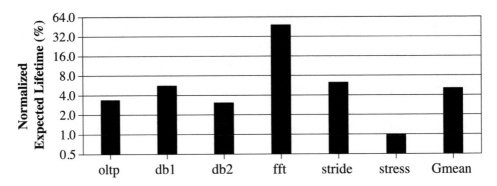

Figure 4.2: Expected lifetime of memory system normalized to uniform-writes.

The lifetime of a PCM system can be increased by making the writes uniform throughout the memory space. *Wear leveling* techniques try to make writes uniform by remapping frequently written lines to less frequently written lines. Existing proposals for wear-leveling [81][116][12][16] [45] use storage tables to track the write counts on a per line basis. The mapping of logical lines to physical lines is changed periodically and the mapping is stored in a separate indirection table. Table based wear-leveling methods require significant hardware overhead (several megabytes) and suffer from increased latency as the indirection table must be consulted on each memory access to obtain the physical location of a given line. This chapter describes simple mechanism that avoids the storage and latency overheads of existing wear-leveling algorithms and still achieves a lifetime close to perfect wear-leveling. The content of this chapter is derived from [154].

4.2   FIGURE OF MERIT FOR EFFECTIVE WEAR LEVELING

The objective of a wear-leveling algorithm is to endure as many writes as possible by making the write traffic uniform. If $Wmax$ is the endurance per line, then a system with perfect wear-leveling would endure a total of ($Wmax \times$ Num Lines In Memory) writes. We define "Normalized Endurance (NE)" as follows:

$$NE = \frac{\text{Total Line Writes Before System Failure}}{\text{Wmax} \times \text{Num Lines In Memory}} \times 100\% \qquad (4.1)$$

Normalized Endurance close to 100% indicates that the wear-leveling algorithm can achieve system lifetime similar to maximum possible lifetime. This metric will be used to report the effectiveness of wear leveling algorithms.

4.3   START-GAP WEAR LEVELING

Existing wear leveling algorithms require large tables to track write counts and to relocate a line in memory to any other location in memory in an unconstrained fashion. The storage and latency overhead of the indirection table in table based wear leveling can be eliminated if instead an algebraic mapping of logical address to physical address is used. Based on this key insight, Qureshi et al. [154] proposed *Start-Gap* wear leveling that uses an algebraic mapping between logical addresses and physical addresses, and avoids tracking per-line write counts. Start-Gap performs wear leveling by periodically moving each line to its neighboring location, regardless of the write traffic to the line. It consists of two registers: *Start* and *Gap*, and an extra memory line *GapLine* to facilitate data movement. *Gap* tracks the number of lines relocated in memory and *Start* keeps track of how many times all the lines in memory have been relocated. Start-Gap algorithm is explained with an example.

4.3.1   DESIGN

Figure 4.3(a) shows a memory system consisting of 16 lines (0-15). To implement Start-Gap, an extra line (*GapLine*) is added at location with address 16. The 17 lines can be visualized as forming a circular buffer. *GapLine* is a memory location that contains no useful data. Two registers, *Start* and *Gap* are also added. *Start* initially points to location 0, and *Gap* always points to the location of the *GapLine*. To perform wear leveling, *Gap* is moved by 1 location once every ψ writes to memory. The move is accomplished simply by copying the content of location of [*Gap*-1] to *GapLine* and decrementing the *Gap* register. This is shown by movement of *Gap* to line 15 in Figure 4.3(b). Similarly, after 8 movements of *Gap* all the lines from 8-15 get shifted by 1, as indicated in Figure 4.3(c).

Figure 4.3(d) shows the case when *Gap* reaches location 0, and Line 0 - Line 15 have each moved by 1 location. As with any circular buffer, in the next movement, *Gap* is moved from location 0 to location 16 as shown in Figure 4.3(e). Note that Figure 4.3(e) is similar to Figure 4.3(a) except that the contents of all lines (Line 0 to Line 15) have shifted by exactly 1 location, and hence the *Start* register is incremented by 1. Every movement of *Gap* provides wear leveling by remapping a line to its neighboring location. For example, a heavily written line may get moved to a nearby read-only line. To aid discussion, the terms *Gap Movement* and *Gap Rotation* are defined as follows:

Gap Movement: This indicates movement of *Gap* by one, as shown in Figure 4.3(a) to Figure 4.3(b). As Gap Movement is performed once every ψ writes to the main memory, where ψ is a parameter that determines the wear leveling frequency. *Gap* register is decremented at every Gap Movement. If *Gap* is 0, then in the next movement it is set to N (the number of locations in memory).

Gap Rotation: This indicates all lines in the memory have performed one Gap Movement for a given value of *Start*. The *Start* register is incremented (modulo number of memory lines) on each Gap Rotation. Thus, for a memory containing N lines, Gap Rotation occurs once every

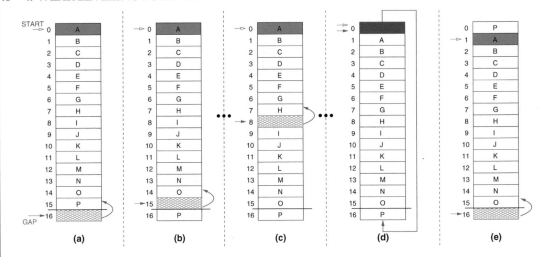

Figure 4.3: Start-Gap wear leveling on a memory containing 16 lines.

$(N + 1)$ Gap Movement. The flowchart for Gap Movement (and Gap Rotation) is described in Figure 4.4.

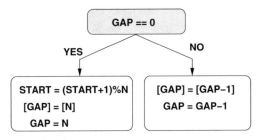

Figure 4.4: Flowchart for Gap Movement.

4.3.2   MAPPING OF ADDRESSES

The *Gap* and *Start* registers change continuously, which changes the mapping of logical to physical memory addresses. The mapping is obtained by making two observations: (1) In Figure 4.3(c) all addresses greater than or equal to *Gap* are moved by 1 and all location less than *Gap* remain unchanged. (2) When *Start* moves as in Figure 4.3(e) all locations have moved by 1, so the value of *Start* must be added to the logical address to obtain physical address. The mapping is captured by the pseudo-code shown in Figure 4.5, which may be trivially implemented in hardware using few gates. If PA is less than N then memory is accessed normally. If PA=N then the spare line (Location 16 in Figure 4.3) is accessed.

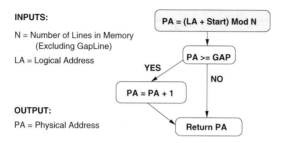

INPUTS:

N = Number of Lines in Memory
(Excluding GapLine)

LA = Logical Address

PA = (LA + Start) Mod N

PA >= GAP

YES

NO

PA = PA + 1

OUTPUT:

PA = Physical Address

Return PA

Figure 4.5: Mapping of Logical Address to Physical Address.

4.3.3   OVERHEADS

A Gap Movement incurs a write (copying data from the line next to *GapLine* to *GapLine*). *Start* and *Gap* must move fast enough to spread hot spots across the entire memory over the expected life time of the memory. However, *Gap* must move slow enough to not incur too many writes. Otherwise, these spurious writes may consume a significant fraction of cell endurance, and would lead to higher power consumption. The frequency of Gap Movement can easily be controlled using the parameter *Gap Write Interval* (ψ). A Gap Movement occurs once every ψ writes. Thus, the extra writes due to wear leveling are limited to $\frac{1}{\psi+1}$ of the total writes. We use $\psi = 100$, which means Gap Movement happens once every 100th write to the memory. Thus, less than 1% of the wear out occurs due to the wear-leveling, and the increase in write traffic and power consumption is also bounded to less than 1%. To implement the effect of $\psi = 100$, one global 7-bit counter can be used that is incremented on every write to memory. When this counter reaches 100, a Gap Movement is initiated and the counter is reset.

The Start-Gap algorithm requires storage for two registers: *Start* and *Gap*, each less than four bytes (given that there are 2^{26} lines in baseline system). Thus, Start-Gap incurs a total storage overhead of less than eight bytes for the entire memory. The *GapLine* can be taken from one of the spare lines in the system. If the memory system does not provision any spare line, a separate 256B line will be required.

4.3.4   RESULTS FOR START-GAP

Figure 4.6 shows the Normalized Endurance for baseline, Start-Gap, and perfect wear leveling (uniform writes). *Gmean* denotes the geometric mean over all six workloads. Workloads db1 (database1), db2 (database2), and oltp are commercial workloads. Start-Gap achieves 20%-60% of the achievable endurance for the three database workloads. The stride kernel writes to every 16th line; therefore, after every 16th Gap Movement, all the writes become uniform and Start-Gap achieves close to perfect endurance. The average endurance with Start-Gap is 53%, which is 10x higher than the baseline.

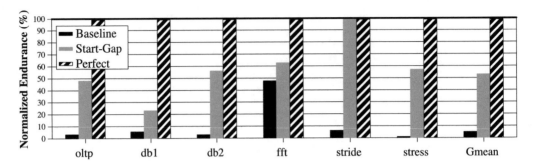

Figure 4.6:   Normalized Endurance with Start-Gap wear leveling with $\psi = 100$.

4.3.5   A SHORTCOMING OF START-GAP

Although Start-Gap improves endurance by 10x compared to the baseline, it is still 2x lower than the ideal. This happens because in each Gap Movement, Start-Gap restricts that a line can be moved only to its neighboring location. If writes are concentrated in a spatially close region, then Start-Gap can move a heavily written line to another heavily written line, which can cause early wear-out. As a counter-example, consider the stride kernel. The heavily written lines are uniformly placed at a distance of 16 from each other. So, Start-Gap is guaranteed to move a heavily written line to 15 lines that are written infrequently before moving it to another heavily written line. Therefore, it is able to achieve close to ideal endurance. Unfortunately, in typical programs heavily written lines tend to be located spatially close to each other, partly because the clock replacement algorithm commonly used in current operating systems searches from spatially nearby pages for allocation.

   Figure 4.7 shows the spatial distribution of writes in the baseline system for db1, fft and stride. To keep the data tractable, the memory is divided into 512 equal regions (128K lines each) and the total writes per region is shown for a period when memory receives 4 Billion writes. Thus, the average writes per region is always 8 Million. For db1, heavily written regions are spatially close between regions 400-460. For fft, about half of the regions are heavily written and are located before region 250. If write traffic can be spread uniformly across regions (like for stride) then Start-Gap can achieve near perfect endurance. In the next section, we present cost-effective techniques to make the write traffic per region uniform.

4.4   RANDOMIZED START-GAP

The spatial correlation in location of heavily written lines can be reduced by using a randomizing function on the address space. Figure 4.8 shows the architecture of *Randomized Start-Gap* algorithm.

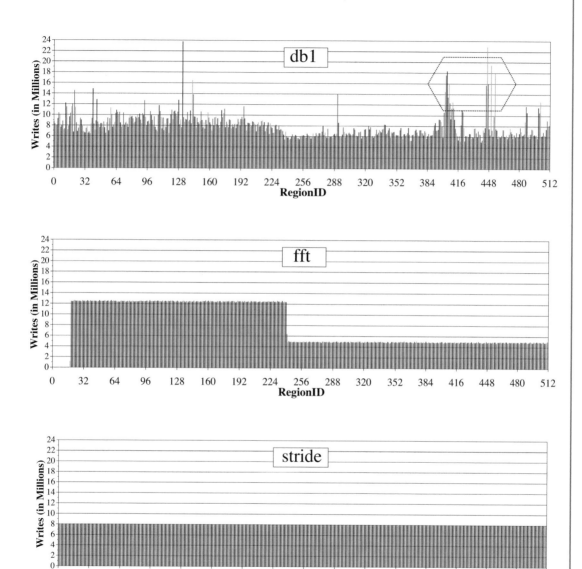

Figure 4.7: Spatial correlation in heavily written lines for three applications. Write traffic is gathered on a per region basis, where region size is 128K lines.

The randomizer provides a (pseudo) random mapping of a given Logical Address (LA) to an Intermediate Address (IA). Due to random assignment of LA to IA, all regions are likely to get a total write traffic very close to the average, and the spatial correlation of heavily written lines among LA is unlikely to be present among IA. Note that this is a hardware only technique and it does not change the virtual to physical mapping generated by the operating system. The Logical Address (LA) used in Figure 4.8 is in fact the address generated after the OS-based translation and Physical Address (PA) is the physical location in PCM-based main memory.

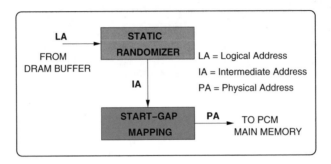

Figure 4.8: Architecture for Randomized Start-Gap.

To ensure correctness, the randomizer must map each line in IA to exactly one line in LA. Thus, the randomizer must be an invertible function. To avoid remapping, a static randomizer is used that keeps the randomized mapping constant throughout program execution. The randomizer logic can be programmed either at design time or at boot time. To be implementable, the randomizing logic must incur low latency and have low hardware overhead. The next sections discuss two such practical designs for randomization.

4.4.1   FEISTEL NETWORK BASED RANDOMIZATION

In cryptography, block ciphers provide a one-to-one mapping from a B-bit plain text to B-bit cipher text. One can use block cipher for randomization. One popular method to build block ciphers is to use the Feistel Network [121]. Feistel networks are simple to implement and are widely used including in the Data Encryption Standard (DES). Figure 4.9 shows the logic for a three stage Feistel network. Each stage splits the B-bit input into two parts (L and R) and provides output which is split into two as well (L' and R'). R' is equal to L. L' is provided by an XOR operation of R and the output of a function (F1) on L and some randomly chosen key (K).

Feistel network has been studied extensively and theoretical work [115] has shown that three stages can be sufficient to make the block cipher a pseudo-random permutation. Qureshi et al. [154] experimentally found that three stages were in fact sufficient for our purpose. The secret keys (*key1, key2, key3*) are randomly generated and are kept constant. For ease of implementation the Function (F1) is chosen to be the squaring function of (L XOR key) as shown in Figure 4.9.

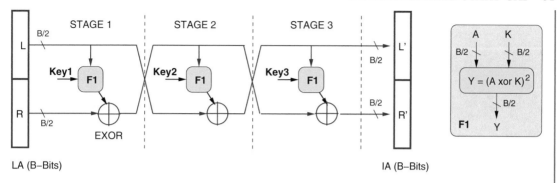

Figure 4.9: Three-stage Feistel Network.

If the memory has B-bit address space ($B = log_2 N$, where N is the number of lines in memory), then each stage of Feistel network requires n-bits ($n = B/2$) bits of storage for the key. The squaring circuit for n-bits requires approximately $1.5 \cdot n^2$ gates [110]. The latency for each stage is $n + 1$ gates [110], which for $B = 26$ is less than 1 cycle even for a very aggressively pipelined processor. Thus, a 3 stage Feistel network would require $1.5B$ bit storage, less than $2 \cdot B^2$ gates, and a delay of 3 cycles.

4.4.2   RANDOM INVERTIBLE BINARY MATRIX

A linear mapping from LA to IA can be performed using a *Random Invertible Binary (RIB)* matrix. The elements of a RIB matrix are populated randomly from {0,1} such that the matrix remains invertible. Figure 4.10 shows the RIB matrix based randomization for an address space of 4 bits. Each bit of IA address is obtained by multiplying one row of RIB with the vector LA. This method uses a binary arithmetic in which addition is the XOR operation and multiplication is the AND operation. Each bit of randomization can be obtained independently (as shown in Figure 4.10 (ii)).

For a memory with B-bit address space ($B = log_2 N$), computing each bit requires B AND gates and *(B-1)* two-input XOR gates. Thus, the total storage overhead of RIB is B^2 bits for matrix, and approximately $2 \cdot B^2$ gates for logic. The latency is delay of $log_2(B)$ logic gates which is less than 1 cycle even for a very aggressively pipelined processor.

4.4.3   RESULTS OF RANDOMIZED START-GAP

Figure 4.11 shows the normalized endurance of the baseline, Start-Gap, and Randomized Start-Gap with RIB-based and Feistel Network based randomization. For the RIB-based scheme, we initialized the matrix with random binary values, ensuring that the matrix remains invertible. For the Feistel Network, the three keys are chosen randomly. There is minor variation (< 1%) in normalized endurance obtained from these schemes depending on the random seed. So, for each workload, the

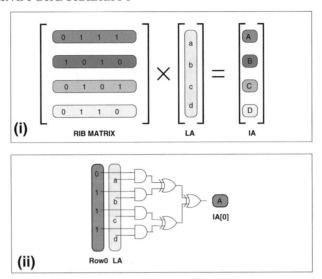

Figure 4.10:   RIB Matrix based randomization (i) concept (ii) circuit for one IA bit.

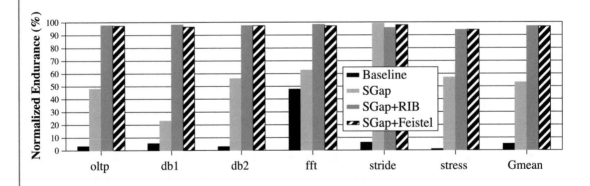

Figure 4.11:   Normalized Endurance of Start-Gap (SGap) with different randomization.

experiments are repeated for both schemes 30 times (each time with a different random seed) and report the average value.

Randomized Start-Gap achieves more than 90% normalized endurance for all workloads. The stride kernel is ideal for Start-Gap as all heavily written lines are equally spaced. Randomizing the address space breaks this uniformity which reduces the endurance slightly. The average across all workloads for Randomized Start-Gap is 97% (with RIB or with Feistel). The total storage required for RIB matrix is 85 bytes and for Feistel network is 5 bytes.[1] So, Randomized Start-Gap requires 93 bytes with RIB and 13 bytes with Feistel network. Thus, Randomized Start-Gap is a practical and effective way to do wear leveling as it achieves near-perfect endurance while incurring negligible hardware overhead. Qureshi et al. also developed an analytical model that explains why Randomized Start-Gap consistently gets lifetime of $>96\%$ for typical programs.

4.5   CONCLUDING REMARKS

The non-uniformity in write traffic across different lines in memory can reduce the effective lifetime of a PCM system significantly. Writes can be made uniform using *Wear leveling*. Unfortunately, existing wear leveling algorithms – developed mainly in the context of Flash – use large storage tables for maintaining logical to physical mapping. These tables incur significant area and latency overhead which make them unsuitable for main memories. In this chapter, we analyze Start-Gap, an algebraic mapping based wear leveling algorithm that obtains near-perfect lifetime for typical applications while requiring negligible hardware and latency overheads. The next chapter describes security considerations for wear leveling.

---

[1] We use B=26, assuming all bits in line address can be randomized. If the memory supports open-page policy then all the lines in the page are required to be spatially contiguous. In that case, only the bits in the address space that form the page address are randomized (randomizing on line granularity or page granularity does not have a significant effect on normalized endurance).

CHAPTER 5

Wear Leveling Under Adversarial Settings

5.1 INTRODUCTION

The last chapter considered wear leveling for typical workloads. However, write limited memories such as PCM and Flash pose a unique security threat. An adversary who knows about the wear leveling algorithm can design an attack that stresses a few lines in memory and cause them to reach the endurance limit, and fail. It is important to address such security loopholes before these technologies can be used in main memories. This chapter describes the possible vulnerabilities, analyze how soon the simplest attack can cause system failure, and discuss solutions that make the system robust to such attacks. The content of this chapter is derived from [154] and [156].

5.2 A SIMPLE ATTACK KERNEL

An adversary can render a memory line unusable by writing to it repeatedly. In a main memory consisting of N lines where there is one gap movement every ψ writes, all the lines will move once every $N \cdot \psi$ writes to memory. For a typical memory system, $N \cdot \psi >> Endurance$, especially for low wear leveling rate (ψ=100). For example, consider a memory bank that has N=2^{24} lines, ψ=100 and Endurance=2^{26}), then the maximum number of writes that can potentially be targeted to a given memory line before it gets moved is approximately 25 times the endurance of that line. Therefore, an attacker can easily cause line failure easily by writing to the same address repeatedly. Static randomization of address space does not help with this attack, because instead of some line (line A), some other line (line B) would becomes unusable. Figure 5.1 shows the pseudo-code of such an attack.

This code causes thrashing in LRU managed cache (assuming address-bits based cache indexing) and causes a write to $W + 1$ lines of PCM in each iteration. The loop can be trivially modified to write to only one line repeatedly in each iteration to result in an even quicker system failure. If the time to write to a given memory line once is one micro second, then a line with Endurance of 64 million will fail in approximately 1 minute.

This code can be run in user space without any special OS support and reduces system lifetime to few hours or days. Given the simplicity and effectiveness of such attacks, it is conceivable that some users may run attack-like codes in their system a few months before the warranty of their machine expires, so as to get a new system before the warranty expiration period. Thus, the security

W = Maximum associativity of any cache in system

S = Size of largest cache in system

```
Do aligned alloc of (W+1) arrays each of size S
while(1){
    for(ii=0; ii<W+1; ii++)
        array[ii].element[0]++;
}
```

Figure 5.1:   Code for attacking few lines.

problem in PCM systems is a concern not only from malicious attackers but also from greedy users who can exploit this vulnerability to potentially obtain a new machine as and when they wish. It is unlikely that any industrial manufacturers will take the risk of designing a main memory system with PCM components unless practical implementations of secure wear leveling algorithms become available.

5.3   SUMMARY OF SECURE WEAR LEVELING ALGORITHMS

Recently several wear leveling algorithms that exploit the fine grained write behavior of PCM have been proposed, e.g., Row Shifting [218], Segment Swapping [218]. It was shown that these algorithms work well for typical workloads. However, these algorithms are deterministic in nature, in that the location of the mapped line can be guessed easily. Therefore, these algorithms are vulnerable to attacks that try to write to the same line repeatedly [154][182][179]. Such attacks are extremely simple to implement and can reduce system lifetime to a few hours under deterministic wear leveling algorithms. A good survey of vulnerability of several recently proposed write filtering techniques and wear leveling techniques can be found in Section 2 of [179]. The acknowledged solution for making wear leveling algorithms secure is to perform remapping in a randomized manner, such that it is very hard for the adversary to guess, after relocation, the physical location of a given line. We analyze three recently proposed security-aware wear leveling algorithms and discuss their RCP.

Figure 5.2(a) provides an overview of Region-Based Start-Gap (RBSG) [154]. RBSG is similar to Start-Gap, except that it contains multiple regions each of which is managed by its own Start and Gap registers. The region size of RBSG is tuned such that the maximum number of potential writes to a line is slightly less than endurance, while maintaining a low rate of one line movement every ψ (100) writes (RCP $= \frac{1}{\psi}$). Unfortunately, such a low rate of movement is insufficient under attacks. A recent study [182] showed that with a low rate of movement, the lifetime of RBSG under malicious write patterns [90] is only a few days. To ensure robustness under attacks, this rate must be increased to more than one line movement every demand write ($>$ 100% overhead).

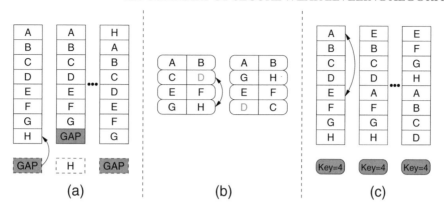

Figure 5.2: Recent Wear leveling algorithms: (a) Region Based Start-Gap (b) Seznec's PCM-S Scheme (c) Security Refresh

Seznec proposed a table-based secure wear leveling scheme *PCM-S* [183], which is shown in Figure 5.2(b). To reduce table size, it splits memory into regions (containing several thousand lines). On a write to a region, the region is swapped with a randomly picked region in memory with a small probability p (RCP = p). The lines within the region are rotated by a random amount (as shown by line D after swap). In this scheme, there is a trade-off between the region size and the overall write overhead. Seznec considered that a write overhead of one extra write per 8 demand writes was acceptable. For 4K-block regions this scheme is able to support malicious attack targeting a single block and still provide about 69 % of the ideal lifetime. In general, on conventional applications even with a much lower p (about 100x lower), the scheme would still provide close to ideal lifetime. However, as proposed, the scheme requires 12.5% write overhead for all applications; this overhead is not gently distributed over time, but in bursts (two mega-byte sized regions must be swapped). While the proposal interleaves the region swaps with the demand writes, the swaps must be performed at a sufficient rate to avoid the need for large buffering; consuming one-fourth of the overall write bandwidth was shown to be effective.

Figure 5.2(c) shows the Security Refresh [179] scheme (single level scheme is shown for simplicity, multi-level scheme is discussed in Section 5.10.3). A random key decides the amount that the entire memory region is rotated. Rotation is performed by swapping two lines. Each swap is performed after every ψ writes (RCP = $\frac{1}{\psi}$). After all lines are rotated, the entire memory shifts by an amount defined by the random key, and then the key value is changed. For the single level scheme, if one swap operation is performed every two demand writes (100% overhead), it obtains 12.5% of normalized lifetime. To obtain higher lifetime under attacks, it incurs > 100% write overhead.[1]

---

[1]If all lines of a region are contained within one bank, and wear leveling is performed at bank level then the analysis of the Single-Level Security Refresh scheme becomes similar to that of randomized Start-Gap.

5.4   FORMULATING SECURE WEAR LEVELING AS BUCKETS-AND-BALLS PROBLEM

The RCP parameter essentially decides the number of writes that can be performed to a given line before that line moves, which determines the vulnerability to attacks that repeatedly write to a given line. We define the *Line Vulnerability Factor* (LVF) as the maximum number of writes to a line before that line gets moved by the wear leveling algorithm. If LVF is close to the Endurance of the line, say Endurance/2, then in each round of attack half the lifetime of line is lost. The attacker just needs to find one of the lines that have been attacked before to cause line failure. This is exploited by birthday paradox attacks, as the expected number of random trials to find a line twice in a set of N lines is only $1.25\sqrt{N}$, which is much lower than N (for example, if N=4Million, one needs only 2.5K trials, 1600x lower). Whereas if the LVF is Endurance/100 then the same physical line must be obtained 100 times in random trials, and one would expect the number of trials to be much closer to theoretical maximum.

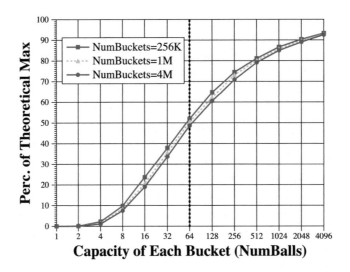

Figure 5.3: Buckets and Balls problem as related to secure wear leveling. Buckets must have a capacity of > 64 balls to get > 50% of theoretical maximum, indicating LVF must be < $E/64$.

A critical question in architecting secure wear-leveling is determining how the obtained lifetime under attack varies with LVF. This can be analyzed using an analogous buckets and balls problem. If there are N buckets, each of which can hold B balls, then under ideal conditions these buckets can hold $N \cdot B$ balls before an overflow of one of the bucket occurs. Now if balls are thrown at random in one of the buckets, then how many balls can be thrown before one of the bucket overflows? Figure 5.3 shows the number of balls thrown (expressed as percentage of theoretical maximum) as the capacity of each bucket is varied from 1 to 4096 balls for N=256K, N=1Million, and N=4Million.

For obtaining more than 50% of the theoretical maximum, the buckets must have capacity of 64 or more balls. Furthermore, this threshold is relatively insensitive to the number of buckets (as the number of buckets is changed from 256K to 4M). In the above formulation, each bucket can be thought of as a line in the bank, and one ball corresponds to one unit of LVF. Thus, the RCP in any given randomized wear leveling algorithm must be tuned such that the LVF is less than Endurance/64. This would mean that under continuous attacks, randomized wear leveling will result in a line failure on average after half the theoretical maximum number of writes are caused to the system. Note that, to keep the analysis general the above calculations did not include the wear due to wear leveling itself (which is a function of the specific implementation of the wear leveling algorithm). Based on this analysis, it is recommended that LVF should be approximately Endurance/128 for a randomized wear leveling algorithm to be secure against repeated write attacks, in order to ensure that the system can last for at-least half the theoretical maximum lifetime even under attacks. Unfortunately, having such a low LVF inevitably results in very high write overhead.

5.5   WRITE OVERHEAD OF SECURE WEAR LEVELING

As each bank operates independently, in our analysis it is assumed that wear leveling is done at a bank granularity (each bank has its own wear leveling algorithm). Assuming a bank size of 4M lines, each of which can endure 64M writes, an LVF of Endurance/128 corresponds to 0.5M writes. To achieve this LVF, Start-Gap would need eight Gap movements per each demand write (800% overhead) and Security Refresh would also need 4 swaps every write (800% overhead). Such a write overhead is clearly prohibitive. Writes in PCM system are power-hungry and slow. Increasing the write traffic would not only increase the wear on PCM cells but also cause significant performance degradation by exacerbating the already scarce write-bandwidth, causing increased read latencies. Furthermore, overall power consumption of PCM systems is predominantly dictated by writes, which means overall energy/power consumption of memory system would increase substantially. Therefore, it is desirable to keep the write overhead to less than 1%. Note that for typical applications a write overhead of less than 1% is sufficient to obtain close to ideal lifetime for these algorithms. Thus, secure wear leveling incurs a write overhead that is 2-3 orders of magnitude overhead than what is required for typical applications. The next section describes a framework that can provide robustness of secure wear-leveling while still incurring negligible overhead for typical applications.

5.6   ADAPTIVE WEAR LEVELING

Secure wear leveling algorithms make common case applications pay a high overhead. A more practical approach to designing a robust wear leveling algorithm is to keep the overheads to minimum for typical applications, and pay a higher overhead for only attack-like applications. Unfortunately, all of the previously described algorithms use a constant RCP which is agnostic of the properties of the memory reference stream. If RCP is low, it makes the system vulnerable to attacks and if the RCP is high then even typical applications pay a much higher overhead for wear leveling. If the wear leveling algorithm is made aware of the properties of the reference streams then it can use higher RCP for only attack like applications and a lower RCP for typical applications. Based on this insight, Qureshi et al. [156] propose *Adaptive Wear Leveling*.

5.6.1 ARCHITECTURE

Figure 5.4 shows the architecture of Adaptive Wear Leveling (AWL). AWL consists of an Online Attack Detector (OAD) that analyzes the memory reference stream to detect attack-like streams. Based on this information, it increases the frequency at which line moves under attack-like scenarios, thus providing more robustness and security.

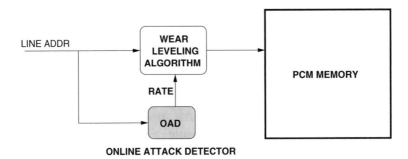

Figure 5.4: Arch. of Adaptive Wear Leveling.

For simplicity, the concept of AWL is explained using Start-Gap as an example (AWL for other algorithms will be explained in Section 5.10). To aid our discussion, a few terms are defined. For Start-Gap, one line moves every ψ writes, for a memory bank with N lines, therefore:

$$\text{LVF} = N \cdot \psi \tag{5.1}$$

If E is the Endurance of the line, and given that the desired value of LVF is $E/128$, therefore:

$$\psi = \frac{E}{128 \cdot N} \tag{5.2}$$

Given that E=64M and N=4M, this would result in a value of ψ much less than one, indicating several gap movements for each demand write to memory. However, it is desirable to have $\psi \geq 100$ to limit the write overhead to less than 1%. To balance these contradictory requirements, our proposal changes ψ from a static value to a dynamic value which depends on the behavior of the memory write stream.

5.6.2   ATTACK DENSITY

In general, between consecutive writes to a given line there are several writes to other locations of memory. Since the system is concerned only about the most frequently written line, the metric of interest is the number of writes between consecutive writes to the most frequently written line. Let *Attack Density* (AD) be the ratio of the number of writes to the most frequently written line to the total number of writes within a given time period. Attack Density is denoted by the symbol Δ. If there are writes to other lines in memory, in between consecutive writes to a given line, this reduces the *Effective LVF* (ELVF) perceived by the line in proportion to Δ:

$$ELVF = LVF \cdot \Delta \tag{5.3}$$

To limit the lifetime lost under one round of attack to $E/128$, the desired value of dynamic ψ, which is denoted by Ψ, is derived as follows:

$$ELVF = \frac{E}{128} \Rightarrow N \cdot \Psi \cdot \Delta = \frac{E}{128} \Rightarrow \Psi = \frac{E}{128 \cdot N \cdot \Delta} \tag{5.4}$$

Thus, Ψ is inversely proportional to Δ, which means that the rate of wear leveling would be directly proportional to Δ. Typically, secure wear leveling algorithms implicitly assume $\Delta = 1$. However, in practice Δ is lower by several orders of magnitude for typical applications as is shown next.

5.6.3   ATTACK DENSITY FOR TYPICAL APPLICATIONS

Given that the last-level cache typically contains more than 100K lines [72], one can expect that for normal applications there will be several thousand writes between consecutive writes to a given line. This was experimentally validated by Qureshi et al. [156] using a 32MB 8-way last-level cache and keeping track of most recent w writes to memory. Table 5.1 shows the fraction of writes that hit in a window of 1K, 2K, 4K writes for six applications: three commercial database workloads (db1, db2, and oltp), fft, stride16 kernel which writes to every 16th line in memory, and stress (8 write-intensive benchmarks from SPEC06).

Table 5.1:   Probability of hit in most recent "w" writes to memory.

	w=1024	w=2048	w=4096
db1	0.67×10^{-6}	1.45×10^{-6}	10.2×10^{-6}
db2	0.01×10^{-6}	0.35×10^{-6}	13.2×10^{-6}
oltp	0.02×10^{-6}	1.19×10^{-6}	24.7×10^{-6}
fft	0.00×10^{-6}	0.00×10^{-6}	00.0×10^{-6}
stride16	0.00×10^{-6}	0.00×10^{-6}	00.0×10^{-6}
stress	0.65×10^{-6}	1.08×10^{-6}	1.21×10^{-6}

If a line is written continuously with less than 1K writes to other lines between consecutive writes then one would expect the probability of hit in these window sizes to be 10^{-3} or higher. However, we find that the probability of hit in a window of 1K writes is less than one in a million, indicating that for typical applications $\Delta \ll 10^{-3}$. Therefore, according to Equation 4, one can safely have $\Psi = 128$ for applications that have $\Delta \leq 10^{-3}$. However, for applications that repeatedly hit within a window of 1K, Δ becomes more than 10^{-3}, and such applications can be deemed as attacks.

5.7   ONLINE ATTACK DETECTION

It is possible to create a hardware circuit that keeps track of most recent 1K write addresses to measure the value of Δ. On each write access, all the recent 1K addresses are checked, and the number of accesses that hit in this window is counted. The value of Δ can be estimated as simply the hit rate of the window, pessimistically assuming all hits are coming from the same line. If the hit count in the window is greater than a certain threshold then the application is likely to be an attack. However, such a circuit would incur impractically large area, power, and latency overhead. A low-cost, practical, yet accurate attack detection circuit can be developed by analyzing some basic attacks.

5.8   ANATOMY OF AN ATTACK

For an attack to successfully cause failure in lifetime limited memories in a short time, it has to write to a *few lines*, *repeatedly*, and at a *sufficiently high write bandwidth*. All the three requirements are important. For example, if the attack simultaneously focuses on several thousand lines, then Δ will be in a range where even the default Start-Gap will move the lines before significant wear-out. The writes must be done repeatedly for several million times for each line; otherwise, the wear-out on each line will be negligible. And, if the attack happens at very low write bandwidth, then the time for the attack to succeed will increase linearly. Figure 5.5 shows canonical form of several attacks. All of these attacks are minor modifications of the repeat address attack, and are extremely simple to implement.

Figure 5.5(i) shows the Repeat Address Attack. It continuously writes to the same line. Therefore, $\Delta = 1$. This attack can be generalized, where the writes are done to n lines continuously. This is called Generalized RAA (GRAA) with period of n, as shown in Figure 5.5(ii). For GRAA, $\Delta = \frac{1}{n}$. Birthday Paradox Attack (BPA) can be viewed as a form of GRAA, which changes the working set after every several million writes. As shown in Figure 5.5(iii), for BPA also $\Delta = \frac{1}{n}$. The final attack shown in Figures 5.5(iv), which is called *Stealth Mode Attack (SMA)*, attacks only one line but disguises it in other $(n-1)$ lines. These lines are chosen randomly and may not repeat across iterations. For SMA again, $\Delta = \frac{1}{n}$, but the attack is concentrated on only 1 line. Figure 5.6 shows the probability of hit in a window of most recent 1K writes for these attacks and typical applications. There is 3-4 orders of magnitude difference between the hit rate from attack-like patterns and

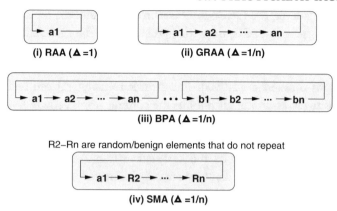

Figure 5.5: Types of attacks (i) Repeat Address Attack (RAA) (ii) Generalized RAA (iii) Birthday Paradox Attack (iv) Stealth Mode Attack.

patterns from typical workloads. SMA is the most challenging to detect among the attacks. If SMA can be detected, then GRAA (hence BPA) can be detected as well, as they have multiple attack lines thus providing a higher chance of being detected. Therefore, it is important that detectors focus on detecting SMA-type attacks.

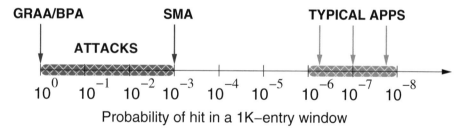

Figure 5.6: Differentiating between attacks and typical applications using hit rate.

5.9   PRACTICAL ATTACK DETECTION

The attack detection circuit must measuring hits in a window of 1K writes, given that in the common case almost none of the line provide a hit. One can approximate the hit rate in such a scenario by having a stack with few entries and simply inserting the address of the incoming write request in the stack with a very small probability p. Such a circuit is called *Practical Attack Detector* (PAD). In addition to the stack entries, PAD also contains two global counters, *HitCounter* and *WriteCounter*, to estimate the hit ratio. Each incoming write address is checked in PAD and increments the

WriteCounter. If there is a hit, the HitCounter is incremented and the replacement information of that line is updated. If there is a miss, then with probability p the given address is inserted in the stack.

If the WriteCounter or the HitCounter reaches its maximum value, the hit rate of the stack is calculated, and both counters are halved. The Attack Density (Δ) can be estimated as simply the hit rate, conservatively assuming all the hits are coming from a single line.[2] For example, for SMA that repeats once every 1K writes, the hit rate will be 10^{-3} and the estimated Δ will be 10^{-3}. The estimated value of Δ is stored in a register *DensityReg* and this value is used for determining the rate of wear leveling between periods of density calculation. A 10-bit HitCounter and a 20-bit WriteCounter is sufficient. Note that since we are only interested in estimating an Attack Density that is higher than 10^{-3}, any density calculation that leads to an estimated $\Delta < 10^{-3}$ is overridden with a value of $\Delta = 10^{-3}$.

A key component of PAD is the replacement policy. Qureshi et al. [156] evaluated both LRU replacement and frequency based replacement and found that frequency based replacement was robust and more accurate than LRU replacement. In general, a 16-entry detector, with insertion probability of 1/256, and 8-bit frequency counters per line is accurate for our purpose.

5.10   IMPLEMENTING ADAPTIVE WEAR LEVELING

The previous sections have discussed circuits that can estimate Attack Density (Δ) accurately. This section discusses how to use this information for implementing Adaptive Wear Leveling (AWL) algorithms. AWL can be implemented in different ways. Figure 5.7 classifies wear leveling paradigms. Recall that, static schemes that always use low rate (Figure 5.7 (a)) are vulnerable to attacks, while always using a high rate incurs high overhead (Figure 5.7 (b)). If accurate estimate of Δ is available throughout the range of Δ then the rate of wear leveling can be increased linearly with Δ (Figure 5.7(c)). However, estimating Δ for extremely low values ($<< 10^{-3}$) incurs significant hardware overhead and complexity; therefore, this is not a practical option. We can define a threshold above which accurate estimates are available. The highest rate is used above this threshold, and lower rate below this threshold (as shown in Figure 5.7(d)). This incurs higher overhead in the range beyond the threshold, as it implicitly assumes $\Delta = 1$ for any point of operation beyond the threshold. Alternatively, a lower rate can be used below the threshold, and linearly increase the rate in proportion to Δ after this threshold. This is shown in Figure 5.7(e), and is the paradigm that will be explained in this section for several recent wear leveling proposals.

5.10.1   ADAPTIVE START GAP

It is desirable that the parameter Ψ of Start-Gap, which determines writes per one Gap movement, be 128 for normal operations (when $\Delta \leq 1/1k$) and that it decrease as Δ increases. Therefore, (Ψ)

---

[2] Assuming that all hits come from a single line leads to a higher estimated value of Δ in multi-line attack. However, this is not a problem as it does not compromise robustness, it only increases write overhead. More importantly, this makes detection of multi-line attacks much easier than single line attacks.

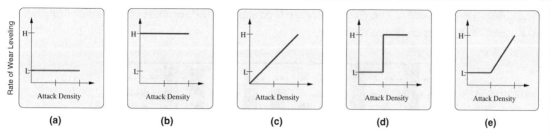

Figure 5.7: Types of Wear Leveling (a) Static - Always low rate (b) Static - Always high rate (c) Adaptive - Linear rate (d) Adaptive - Step function rate (e) Adaptive - Piecewise linear rate).

can be calculated as follows:

$$\Psi = \min\{128, \frac{1}{8 \cdot \Delta}\} \tag{5.5}$$

Thus, Adaptive Start-Gap (ASG) will need to perform 8 gap movements for each demand write for an attack with $\Delta = 1$. This is very high write overhead, albeit under attacks performance is a secondary concern. Still, it is desirable to limit the write overhead to no more than 100% (one extra write per demand write). The overhead of $> 100\%$ overhead can be avoided by leveraging the write queue associated with each bank [157] and employing the delayed write policy (DWP) [154]. If the estimated $\Delta \geq 1/8$ then the write queue enforces DWP (not writing unless there are 8 entries) to ensure that the attacker has to write more than 8 lines for the line to be actually written. Thus, even the worst-case overhead would be limited to one extra write per demand write. Note that extra storage is not needed to implement DWP, as PCM-based systems are likely to contain large write queues for performance reasons [157].

Lifetime Under Attacks: For evaluation in this section it is assumed NumLinesInBank = 2^{22}, Endurance = 2^{26}, and TimeToWriteOneLine = 1μsec. The manufacturer may rate the memory lifetime under the scenario that addresses are written sequentially at full speed. This means, an achievable lifetime of 8 years for each bank[3] of memory and the memory would be rated for eight years. Figure 5.8 shows the rated lifetime and lifetime under different attacks. Attacks with a period of less than 8 are satisfied by the write queue and provide virtually unlimited lifetime, and hence are not meaningful. For GRAA(period=8) ASG get lifetime of 8 years because ASG converts this attacking pattern into sequential write pattern. The overhead of ASG is one extra write per demand write, which slows the system and delays the success of attack. For Birthday Paradox Attacks (BPA) with period of 8, the achievable lifetime is 5.5 years. As more and more lines are attacked the lifetime under BPA increases and the overhead decreases. At 1K and beyond, ASG gets > 7 years of lifetime and the overhead starts to become negligible. Similarly, with GSMA the lifetime is 7+ years with

---

[3] The number of lines per bank is 2^{22}, each line can be written 2^{26} times, for a total of 2^{48} writes per bank. If 2^{20} writes per second can be performed for each bank, with approximately 2^{25} seconds per year, it would take 2^{3} years.

negligible write overhead. Thus, Adaptive Start-Gap achieves 65% or more of the achievable lifetime of 8 years, even under continuous attacks.

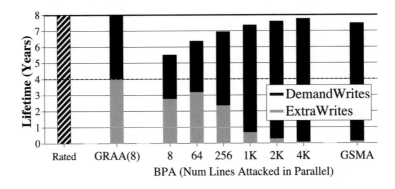

Figure 5.8: Lifetime of ASG under attacks.

Table 5.2: Lifetime and Overhead with Adaptive Start-Gap for typical applications.

	db1	db2	oltp	fft	stride16	stress	Average
Norm. Lifetime	98%	98%	97%	98%	96%	93%	97%
Write Overhead	1/128	1/128	1/128	1/128	1/128	1/128	1/128

Lifetime For Typical Applications: For typical applications there are negligible hits in the attack detector and the rate of wear-leveling defaults to once every 128 writes. Table 5.2 shows the lifetime normalized to perfect wear leveling and the write overhead of ASG for the six workloads considered in our study. Unlike attacks, these workloads do not write to memory continuously; therefore, their theoretical maximum lifetime under perfect wear-leveling is much higher than 8 years, and is in the range of 10-40 years. For these workloads the ASG scheme achieves 97% of theoretical maximum lifetime of 10+ years, which is well beyond typical server lifetime [118], while incurring a negligible write overhead ($< 1\%$). Thus, Adaptive Start-Gap is both robust (years of lifetime under attacks) and practical (negligible overhead for typical applications).

5.10.2 ADAPTIVE SECURITY REFRESH (SR-1)

For the purpose of our analysis, Single Level Security Refresh (SR-1) scheme is similar to Start-Gap. In SR-1, one swap is performed every $\lambda = 2 \cdot \Psi$ writes. A swap operation incurs two writes, instead of the one write incurred by the Gap movement in Start-Gap. To obtain an overhead of 1/128 for typical applications, the rate control parameter of SR-1 must be selected such that:

$$\lambda = 2 \cdot \Psi = \min\{256, \frac{1}{4 \cdot \Delta}\} \tag{5.6}$$

The overhead and effectiveness of adaptive version of SR-1 is similar to Adaptive Start-Gap. It gives 5.5 years of lifetime under attacks (while incurring a maximum write overhead of 100%). For typical applications, the achieved lifetime is approximately 97% of theoretical maximum and the write overhead reduces to 1/128.

5.10.3 ADAPTIVE SECURITY REFRESH (SR-M)

The robustness of Single Level Security Refresh (SR-1) can be improved by splitting the memory into multiple regions, where each region performs remapping within the region at an inner rate (IR) and regions are swapped at an outer rate (OR). This is the *Multi-Level Security Refresh* (SR-M) scheme [179]. To ensure more than 50% lifetime for Endurance=64M, the two rates must be set to approximately 1/64. Therefore, the write overhead of SR-M is 3.25%. The write overhead of SR-M can be reduced by controlling both IR and OR. The IR rate can be controlled with a single-entry per-region PAD, and the OR can be controlled by a global PAD. As both IR and OR can be reduced by 16x, the overall write overhead would reduce by a factor of 16, down from 1/32 (3.25%) to 1/512 (0.2%).

Table 5.3: Overhead of Security Refresh Schemes: with and without adaptation.

Scheme	Storage	Write Overhead	
		Attacks	Typical apps.
SR-1	12 bytes	100%	100%
Adaptive SR-1	80 bytes	100%	1/128
SR-M	12 KB	1/32	1/32
Adaptive SR-M	15 KB	1/32	1/512

Table 5.3 compares the hardware overhead and write overhead of SR-1, Adaptive SR-1, SR-M and Adaptive SR-M schemes. Adaptive SR-1 has both lower storage overhead and lower write overhead (for typical applications) than SR-M. Nonetheless, adaptation can be added to SR-M as well to reduce its write overhead. It is noteworthy that the write overhead of SR-M is strongly a function of endurance and increases significantly at lower endurance. Adaptive SR-M is particularly attractive at lower endurance, where SR-M can incur significant write overhead for all applications.

5.11 CONCLUDING REMARKS

A robust wear leveling algorithm must ensure that PCM-based memory system remains durable even under the most stressful write streams. Recently proposed secure wear leveling schemes that are robust to attacks perform memory remapping at very high rate assuming that memory is always under attack. Therefore, they incur very high write overhead, which makes them unattractive for practical implementation. Ideally, we would like to have low write overhead in common case while still being robust to attacks. In this chapter we discussed a more practical approach to secure wear leveling based on the concept of online detection of stressful write patterns. While we used online attack detection for implementing adaptive wear leveling, the general concept can be used for initiating other preventive actions as well. For example, identifying malicious applications and reporting them to OS, and monitoring quality of write stream throughout machine lifetime.

CHAPTER 6

Error Resilience in Phase Change Memories

6.1 INTRODUCTION

One of the major impediment in using PCM for main memories is limited write endurance, currently projected to be in the order of few tens of millions. Even with this limited endurance, it is possible to make a memory system last for several years with effective wear leveling if all cells in the memory array can be guaranteed for that range of write endurance. Unfortunately, process variation can cause a high variability in lifetime of different cells of the same PCM array. Thus, a very small percentage of weak cells (cells that have endurance much lower than average cells) can limit the overall system lifetime. Ensuring long lifetimes in such scenarios requires that the design can either tolerate (through graceful degradation) or correct (via coding techniques) a large number of errors for any given memory line or page. This chapter describes the basic fault model for PCM and some of the recent approaches to make system resilient to these faults. The content of Section 6.2 and Section 6.3 is derived from [66], Section 6.4 is derived from [175], Section 6.5 is derived from [180], and Section 6.6 is derived from [214].

6.2 FAULT MODEL ASSUMPTIONS

PCM cells are not susceptible to radiation induced transient faults in the foreseeable future, due to the high energy required to change their state. The major fault model for PCM cells is permanent failure of a cell.[1] The heating and cooling process required to write a cell, and the expansion and contraction that results, eventually cause the heating element to detach from the chalcogenide material making it impervious to programming pulse. Thus, a popular way to model the write induced faults in PCM is to capture them as a stuck-at fault. Fortunately, the content of such stuck cells can still be read reliably, which helps in detecting such stuck-at faults by performing a checker read following the write operation [66].

It is also reported in [66] that such permanent faults shows no large systematic effects are observed within a die, and in general, if such correlations were observed, they would likely be artifacts of such positional effects as wordline and bitline location within the array, which would be compensated and trimmed using additional circuitry as standard practice. This assumption allows

---

[1]The peripheral circuits in PCM array are still susceptible to soft errors due to radiation effects. Furthermore, PCM cells may have transient failures due to other reasons, such as resistance drift or read disturb failures. In this chapter we assume that the dominant failure mode for PCM is endurance related hard faults and focus only on hard error correction techniques.

the cell lifetime to be modeled using a normal distribution with a given coefficient of variation (CoV). Typical values of CoV assumed in recent studies ranges from 10% to 30% of the mean, with mean writes of 10^8.

6.3 DYNAMICALLY REPLICATED MEMORY

The lifetime of PCM system can be improved under high varibility by providing error correction. Unfortunately, existing error correction schemes for DRAM and NAND Flash are not appropriate for PCM. For example, pages protected with traditional ECC are decommissioned as soon as there is first uncorrectable error in the page. This approach may be acceptable if probability of failure is very low and errors accrue only near the end of the device's life. However, if failures happen early in lifetime due to process variations, this overly conservative strategy can decommission large portions of memory quickly. To tolerate permanent data errors seamlessly, Ipek et al. [66] proposed *Dynamically Replicated Memory (DRM)*, a new technique that allows for graceful degradation of PCM capacity when hard failures occur.

6.3.1 STRUCTURE

DRM is based on the idea of replicating a single physical page over two faulty, otherwise unusable PCM pages; so long as there is no byte position that is faulty in both pages, every byte of physical memory can be served by at least one of the two replicas. Redundant pairs are formed and dismantled at run time to accommodate failures gracefully as faults accrue. The key idea is that even when pages have many tens or even hundreds of bit failures, the probability of finding two compatible real pages, reclaiming otherwise decommissioned memory space remains high.

To facilitate dynamic replication, DRM introduces a new level of indirection between the system's physical address space and PCM in the form of a real address space, as shown in Figure 6.1. Each page in the physical address space is mapped to either one pristine real page with no faults, or to two faulty but compatible real pages that have no failures in the same byte position, and can thus be paired up to permit reads and writes to every location in the corresponding physical page.

6.3.2 PAGE COMPATIBILITY

Figure 6.2 illustrates the concept of page compatibility. It shows two pairs of real pages, one compatible and the other incompatible. In each pair, the dark colored page (P0 or P2) represents the primary copy, and the light colored page (P1 or P3) is a backup page. In the figure, pages P0 and P1 are compatible since there is no byte that is faulty in both of them. If a single physical page were to be replicated on P0 and P1, P0 could serve requests to bytes B0 and B7, P1 could satisfy accesses to B4, and accesses to all other bytes could be served by either page; in this way, a single physical page could be represented by and reconstructed from two faulty, unreliable real pages. Unlike P0 and P1, however, pages P2 and P3 both have a fault in byte position B4, and are thus incompatible: if P2

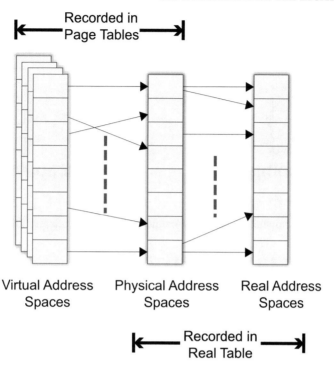

Recorded in Page Tables

Virtual Address Spaces   Physical Address Spaces   Real Address Spaces

Recorded in Real Table

Figure 6.1: Physical Address Space with Dynamic Replication. (Figure adapted from [66].)

and P3 were paired and a physical page mapped onto them, it would not be possible to read or write the data in B4.

6.3.3   ERROR DETECTION

DRM detects errors by performing a checker read after each write to determine whether the write succeeded. The key observation that makes such read-after-write approach viable is that the write operation are much more expensive (in terms of both latency and energy) compared to read operation in PCM. Therefore, if a checker read is issued after each array write to PCM, the read latency gets amortized over the much larger write latency, with only a small effect on overall system performance.

Once a location has been marked as faulty, the memory controller attempts to write the data to the backup location for that page, if it exists. If the backup page fails, then it too is marked as faulty and then a recovery mechanism is invoked for performing error recovery via relocation.

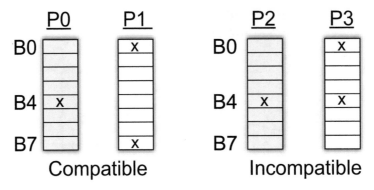

Figure 6.2: Example of compatible and incompatible pages. (Figure adapted from [66].)

6.3.4 LOW OVERHEAD APPROXIMATE PAIRING

When a write fails on both locations in a given pair, or on any location in a pristine page, the memory controller must detect the fault via the checker read and initiate recovery by copying the data to a new location in PCM. Matches between two pages are determined by issuing reads to both pages, computing parity to determine which bytes are faulty, and then comparing the results from both pages to determine whether they have faults at the same location. These comparisons are the most expensive portion of the matching algorithm

Therefore, an approximate pairing algorithm is evaluated, which is both incremental and local. Whenever a page is under consideration for pairing, it is compared to the pages currently residing in the unmatched list one by one. As soon as a compatible page is found, the two pages are paired. The approximate pairing algorithm was found to be very effective for up to 160 errors per page for a pool of 50k pages.

6.4 ERROR CORRECTING POINTERS

As DRM manages faults on a per-page basis, and performs correction by employing two-way redundancy at the page level, it results in high overhead. The overheads to handle faulty pages can be reduced by providing error correction on a per line basis. Unfortunately, the standard ECC implementations used in DRAM are less than ideal given these three aspects of PCM devices: the strong need to reduce writes, the dominance of hard failures, and the ability to identify failures at the time of the write. Hamming-based ECC codes modify the error-correction bits whenever any data in the protected block change, resulting in high entropy that increases wear.

Traditional Error-Correcting Codes (ECC) store sufficient information to derive the location of the error at encoding time, allowing them to correct soft errors discovered when a block of data is read. Conversely, Error-Correcting Pointers (ECP), proposed by Schechter et al. [175], logs the position of the erroneous bit and the correct value. ECP operates within each memory chip at the row level. Each correction unit of ECP consists of a pointer and a replacement bit. The ECPn scheme uses n such correction units to tolerate n bits of hard faults. Figure 6.3 (a) shows the simplest ECP implementation, ECP1, where a single correction entry corrects up to one bit. The example uses a row with 512 data bits. When no errors are present in the data, the correction pointer is empty, and the full bit is set to 0 (false). This indicates the entry is inactive as there are no errors to correct.

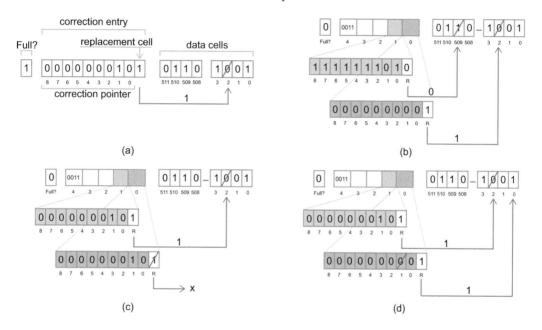

Figure 6.3:   Error Correction with ECP replacement of failed memory cells. (a) Correcting one bit error with ECP-1. (b) Correcting up to five errors with ECP-5. (c) Tolerating errors in replacement cells. (d) Tolerating errors in pointers of ECP (Figure adapted from [175].)

When a bit fails, for example bit 2 in Figure 6.3(a), the correction entry is marked full (or active), the correction pointer is set to point to bit 2, and the replacement cell now stores the value that belongs in bit 2. Henceforth, when the row is written, the value to be written to the failed cell identified by the correction pointer is instead written to the replacement cell. When the row is read, the value in the replacement cell supersedes the value read from the defective cell.

ECP1 can be generalized to an arbitrary number of entries (ECPn) as shown in Figure 6.3(b), using ECP5 as an example. The full bit is now set only when all error- correction entries are in use. Otherwise, the full bit is set to 0 and the bits of the last correction entry contain a unary-encoded counter denoting how many of the other active.

Errors in replacement cells are less likely to occur than errors in the original data cells as they do not begin to wear until they are put into use to replace a failed cell. ECP can still correct these errors as well. When two correction entries point to the same cell, the correction entry at the higher index takes precedence over the one with the lower index. For example, in Figure 6.3 (c), the replacement cell in correction entry 0 has failed. To compensate for this, a correction entry is activated and it points to the same failed cell. The replacement cell in entry 1 supplants both the original failed cell and the failed replacement cell in entry 0. Precedence rules also make possible the correction of errors in correction pointers.

An error in a correction pointer, as illustrated in Figure 6.3(d), effectively replaces a working cell with another working replacement cell, doing no harm. Thus, ECP can correct errors both in data cells and in its own data structures, while allocating only enough bits per correction pointer to address data bits. ECPn can correct any n cell failures, regardless of whether they are in data cells or correction entries. ECP requires 1 full bit, n replacement bits, and n pointers large enough to address the original data bits. Thus, the fractional space overhead of ECPn for a row with d = 512 data bits is 11.9%. Lifetime evaluations show that ECP is more efficient at correcting hard errors than ECC while incurring similar storage overhead.

6.5   ALTERNATE DATA RETRY AND SAFER

ECP uses a 9-bit pointer plus one data bit, so a total of 10 bits, for correcting one bit of error. If a single bit error can be corrected with fewer bits, it would proportionately reduce the storage overhead. Fortunately, ideas for efficiently handling single bit stuck-at-fault exists in literature since at-least the 1970s. One such idea, called *Alternate Data Retry (ADR)* [185], leverages the fundamental insight that a stuck-at-fault cell can still be read. If a stuck-at-fault is detected on the first try, ADR simply masks that fault by rewriting the data in inverted form. Thus, ADR can correct single-bit hard faults by incurring an overhead of just one bit (to denote if the data is inverted or not).

Extending the idea of ADR to handle multi-bit faults is non trivial, and the associated storage overhead increases significantly. Seong et al. [180] proposed SAFER, a dynamic data block partitioning technique that ensures each partition has at most one failed bit; thereby, enabling recovery per partition using storage-efficient ADR technique.

Figure 6.4 explains how SAFER can provide double error correction (DEC) for a data block of 8 bits. Each bit in the data field can be identified with a 3-bit pointer as shown in Figure 6.4(a). If the data block has only one error, as shown in Figure 6.4(b), then any bit position can be used to partition the data-bits in two groups such that each group has at most one error. Lets say position 3 has a data error. Now if another location, say position 0, also has an error then the most significant bit cannot be used to partition the data bits (as both position 0 and 3 have a common MSB). That leaves us with two choices. If the first LSB is chosen, the resulting partition is shown in Figure 6.4(c)(1), and instead if the second LSB is chosen, the resulting partition is shown in Figure 6.4(c)(2).

For a n bit data block, the partition field uses only $log_2(log_2 n)$ additional bits to identify how a block is partitioned. Thus, SAFER is much more efficient that ECP, which requires $log_2 n$ bits. The concept of SAFER can be used to partition blocks with more than two erroneous bits as well. Lifetime evaluations suggests that SAFER can obtain lifetime similar to ECP, while incurring 55 bits/line overhead as opposed to 61 bits/line for ECP.

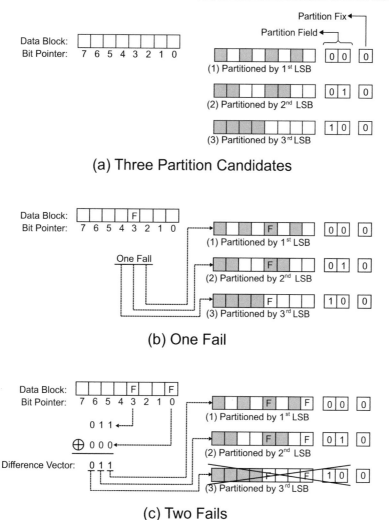

Figure 6.4: Example of partition two faults with SAFER. (Figure adapted from [180].)

Unfortunately, both ADR and SAFER require extra writes to perform error correction, which incurs power, performance, and wear out penalties. The additional writes can be avoided by using a fail cache to identify the positions of the failed bit, and morph the data to mask the hard fault on the first write itself. However, this incurs additional design and area overheads [180].

6.6 FINE-GRAINED EMBEDDED REDIRECTION

Previously discussed error correction techniques focused mainly on handling only hard errors in PCM. However, PCM is susceptible to new forms of transient errors as well, primarily due to resistance drift. Furthermore, those techniques require extra storage to support chipkill that enables a memory DIMM to function even when a device fails. Doe et al. [214] proposed *Fine-grained Remapping with ECC and Embedded-Pointers (FREE-p)* which integrates both soft error and hard error tolerance, and is also amenable to chipkill.

The central theme in FREE-p is fine-grained remapping (FR) of faulty blocks to a reserved area, thereby keeping the page still usable even if the page contains some faulty lines. Without FR, a single cache line failure will deactivate an entire page, leading to poor efficiency. Thus, FR both increases the lifetime of a page and makes the memory more tolerant to process variation. Furthermore, because FR operates at line granularity, it can easily tolerate imperfections in wear-leveling. The operating system is responsible for identifying a remap region for the failed block. In this way, the size of memory degrades gracefully with age.

FR naturally uses a remap granularity of a cache line. This means a naive implementation of FR can incur significant storage overhead as any line could potentially be mapped to the reserved area. Such a high storage overhead would be impractical to keep in the processor caches or in main memory. FREE-p avoids the remap table by leveraging the key insight that even if a block has more faults that can be handled at a per-line granularity, it still has enough bits to store a redirection address. Therefore, FREE-p uses the remaining working bits in the faulty block to embed a remapping pointer, as shown in Figure 6.5.

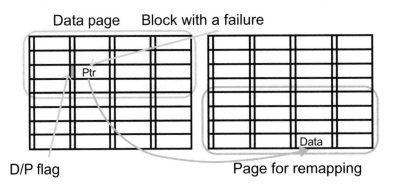

Figure 6.5: Example of fine-grained remapping with embedded pointer. (Figure adapted from [214].)

To identify if the block stores a redirection address or data, an extra bit per line is needed, which is called the D/P flag. This bit is initialized to zero when the line stores data. When a wear out failure is detected, the faulty block is remapped to a fault-free location, and D/P flag associated with the line is set, and the redirection address is stored. Given that this address is only a few bits, and that this information must be stored reliably in the faulty line, a heavy 7 modulo redundancy code is used.

When the remapped line fails too, the scheme can be applied to relocate the line to another line. Instead of following the chain of pointers, FREE-p simply updates the pointer in the original line to point to the current location of the line.

6.7   CONCLUDING REMARKS

Recent studies have indicated that the dominant failure mode in PCM memory system is likely to be hard failures due to variance in lifetime of different cells of a PCM array. To tolerate such lifetime failures, this chapter described several techniques, including dynamic memory replication, write-efficient coding, and embedded redirection of failed lines. While the focus of past research has been dealing with endurance-related hard faults, PCM cells are also susceptible to newer failure modes such as due to resistance drift and read disturb. Therefore, future research can investigate storage-efficient techniques that are tolerant to both hard and soft errors, taking into account the specific failure modes of PCM.

CHAPTER 7

Storage and System Design With Emerging Non-Volatile Memories

7.1 INTRODUCTION AND CHAPTER OVERVIEW

In this Synthesis Lecture thus far, we have presented the characteristics of several key Non-Volatile Memory technologies, made the case for PCM as the leading contender for large scale production in the near future, and then discussed architecture-level techniques to address the performance, reliability, and energy-efficiency challenges that PCM poses when used in main memory systems. We now discuss the utility and broader implications of using PCM and PCM-like Non-Volatile Memory at the systems level, with case studies from the literature.

This chapter is divided into two parts. The first part delves into the design of storage systems that use Non-Volatile Memory. Since modern storage systems already use one type of Non-Volatile Memory - NAND Flash - we first differentiate Flash memory from those NVMs that are similar to PCM. We define an abstraction called *Storage Class Memory (SCM)* that captures the key attributes of the latter class of NVMs. We then provide a brief overview of Flash-based Solid-State Disks (SSDs), highlight certain key challenges in managing such SSDs, and then discuss key ideas from the literature on addressing these challenges through the use of SCMs. The literature on the use of SCMs in systems design span several memory technologies (FeRAM, PCM, STT-RAM, etc.) and diverse application scenarios, from embedded systems to server storage. Therefore, in order to use the SCM abstraction in our discussions, we do not provide quantitative data on the performance, power consumption, and other figures of merit. Instead, we generalize and summarize the key ideas and insights from these papers. The interested reader is encouraged to read the cited papers to gain further understanding about the context of each individual work. The storage systems part of the chapter concludes with a case study of the *Onyx* PCM-based storage array prototype that was built at the University of California, San Diego [5].

The second part of this chapter addresses the implications of memory system non-volatility on systems design. We discuss three different scenarios where the non-volatility property of SCMs can affect system design, with one case study of each from the literature. The first case study we discuss deals with file system design when main memory is non-volatile. The second case study relates to the design of persistent object stores for SCM-based main memory. The final case study relates to non-volatility as a design knob in hardware, where retention time can be traded off for improvements in performance and power.

7.2   STORAGE-CLASS MEMORY - A SYSTEM LEVEL ABSTRACTION FOR PHASE CHANGE MEMORY

Chapter 1 presented a survey of several Non-Volatile Memory technologies. Of these, NAND Flash memory is widely used today in Solid-State Disks (SSDs). As Section 1.3 highlighted, Flash memory faces several scalability, performance, and reliability challenges. Accessing a NAND Flash memory array (shown in Figure 1.4) also has several idiosyncrasies. Reads and write to NAND Flash are done at the granularity of an entire wordline worth of cells (*Flash Page*), which typically tend to be a few Kilobytes in size. In order to write into a Flash page, the cells in the page need to be in the *erased* state. As a result of this property, NAND Flash does not support in-place writes to data. Furthermore, an erase operation can only be performed on a *block* of pages.

While the other NVMs discussed in Chapter 1 and NAND Flash share the property of non-volatility, the others have two key characteristics that Flash does not have, namely:

1. They are byte-addressable.

2. They support in-place writes.

In this Chapter, we refer to Non-Volatile Memory that have the above two properties as *Storage Class Memory (SCM)*. Therefore, while both PCM and Flash memory are NVMs, only PCM is considered an SCM. While individual SCM technologies may vary significantly in terms of read and write latencies, endurance, etc. (as summarized in Table 1.2), we will use this SCM abstraction in the discussions that follow and assume that these costs can be addressed at lower levels of the design stack. However, note that evaluating metrics such as performance and energy-efficiency will require careful consideration of these technology-specific costs.

7.3   STORAGE SYSTEM DESIGN

Storage constitutes a key component of any computing system. The storage system needs to provide sufficient capacity, deliver high performance, and meet these goals in an energy-efficient manner. The energy-efficiency of the storage system is paramount across the spectrum of computing environments, from mobile handheld devices to entire data centers. The advent of Solid-State Disks (SSDs) has played an important role towards realizing such high-performance, energy-efficient storage systems.

7.3.1   OVERVIEW OF SOLID-STATE DISKS

An SSD is a device that uses a Non-Volatile Memory as the storage medium. Current SSDs use NAND Flash as the NVM. Unlike a Hard Disk Drive (HDD), which is an electro-mechanical device with a rotating spindle and moving arms, an SSD has no moving parts. As a result, an SSD typically dissipates much less power and heat than an HDD, makes no acoustic noise, and is tolerant to shock and vibrations. SSDs also use traditional storage interfaces, such as SCSI and SATA, and can therefore be used as drop-in replacements for HDDs.

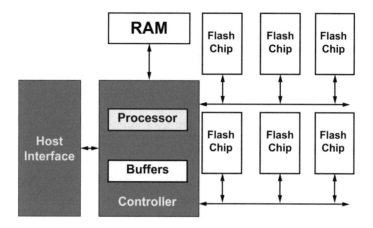

Figure 7.1: Solid-State Disk Architecture. Image Adapted from [4].

The overall architecture of an SSD is given in Figure 7.1. The SSD consists of several Flash memory chips that are connected to a controller. An enterprise SSD may contain more than one controller. The SSD also contains DRAM memory that is used a cache and the storage interface (e.g., SATA). A detailed discussion on the architecture of SSDs is given in [4] and [27].

7.3.2   THE FLASH TRANSLATION LAYER (FTL)

The idiosyncrasies of Flash memory, such as its inability to do efficient in-place writes, break the conventional block-based device abstraction that rotating magnetic disk drives have provided to the other layers of the system stack. Furthermore, there is an asymmetry in the read, write, and erase latencies to Flash memory that further complicates Flash memory management. (The typical read, write, and erase latencies of an SLC Flash chip are approximately 25 microseconds, 200 microseconds, and 1.5 milliseconds, respectively [170]). Since one of the key requirements of SSDs has been to support standard storage interfaces and therefore serve as drop-replacements for HDDs, it is important to handle these Flash-specific behaviors behind the storage interface. This is done via a *Flash Translation Layer (FTL)* within the SSD. The key components of an FTL include:

- **Logical Block Map:** Since each write to a Logical Block Address (LBA) of the disk results in a different Flash memory page being written, we need to translate the LBA to the appropriate physical Flash page. This translation is provided by means of a mapping table. This mapping table is stored in the DRAM memory of the SSD to provide efficient read/write access and is reconstructed at system startup [4]. The target Flash memory page for each LBA write is selected from an allocation pool of free pages that is maintained by the FTL.

- **Garbage Collection:** Due to the copy-on-write nature of writes to Flash memory, each write to an LBA leaves behind a older version of the Flash page that is no longer valid. In order to prevent such stale pages from reducing the available capacity of the SSD, the FTL invokes garbage collection to erase the blocks containing such stale pages and add those blocks to the allocation pool. Since Flash memory blocks may contain a mix of stale and valid pages, the valid pages may first have to be moved to blocks that are already in the erased state as part of the garbage collection process.

- **Wear Leveling:** Flash memory cells have limited endurance due to the write and erase operations. This limited endurance is primarily due to the trapping of charges in the tunnel oxide of the Flash memory cell. Due to this limited endurance, the FTL distributes the write and erase operations evenly over all the blocks in the SSD using a wear leveling algorithm. One of the key design issues in the FTL is to strike a balance between efficient garbage collection, which can induce several write and erase operations on Flash, and wear leveling, whose goal is to extend the lifetime of the memory cells. A detailed survey of wear leveling algorithms for Flash memory and techniques that balance wear leveling with garbage collection is given in [45].

7.4   FTL AND SSD DESIGN OPTIMIZATIONS WITH STORAGE-CLASS MEMORY

The use of an SCM in lieu of Flash memory offers several advantages. First, the ability to perform efficient in-place writes can greatly simplify the design of the FTL by eliminating the need for the Logical Block Map and garbage collection. This property of SCMs can also simplify the storage requirements for implementing wear leveling [154]. The higher write endurance of certain SCMs, such as PCM and STT-RAM, can reduce the need for implementing complex wear-leveling algorithms and boost overall reliability. Finally, the lower latency of certain SCMs (e.g., PCM) can boost overall storage performance compared to Flash-based SSDs.

There are two choices for incorporating an SCM into an SSD: (1) replace all the Flash memory in the SSD with SCM; (2) build an SSD that has both SCM and Flash memory. While both approaches are technically valid, replacing all the Flash with an SCM mandates that the cost-per-Gigabyte of the SCM be low enough to be competitive with regard to both performance and capacity. Given past trends in memory and storage technologies, it is unlikely that an SCM will reach the commodity price point soon after it is introduced into the market. A more likely scenario is for a Flash-based SSD to be augmented with a smaller amount of SCM (to minimize costs), similar to how NAND Flash was initially used in conjunction with rotating media as "hybrid" drives before the price-per-Gigabyte of Flash became competitive enough for it to be used as the sole storage medium within a drive.

In the discussion that follows, we will first discuss how augmenting an SSD with an SCM can benefit storage design and discuss how such a change to the architecture can impact storage interfaces and the software stack. We will consider the case of an SCM-only SSD in Section 7.4.3.

7.4.1   ADDRESSING THE SMALL-WRITE PROBLEM OF FLASH SSDS

Since writes to Flash are generally performed at the granularity of entire pages, writing data that is smaller than a page results in inefficient use of storage capacity. Furthermore, the slow write speeds and the inability to perform efficient in-place writes to Flash can seriously degrade the performance and reliability of SSDs. Such frequent small-writes are the result of operations such as metadata and log-file updates and are common data patterns in many enterprise applications. The byte-addressability, lower latency, and the in-place update capabilities of SCM can be leveraged to address the small write problem in SSDs.

There have been prior studies into augmenting a Flash-based SSD with a small amount of SCM to optimize file metadata access. Doh et al. [38] proposed an architecture called *Metadata in NVRAM (MiNV)*, where file metadata is stored on the SCM and the file data on Flash. In order to use such an SSD, they designed a custom file system called *MiNVFS* and demonstrated performance gains and a reduction in the number of Flash erase operations. Similarly, Kim et al. [84] proposed to use the SCM to store code and file metadata and use Flash for file data storage and proposed an FTL called *Hybrid FTL (hFTL)* to manage the space. While all these approaches leverage the benefits of SCM, they require modifications to the file system and/or the block-device interface to separate the metadata accesses on the SCM and the file data accesses on Flash.

Smullen et al. [191] proposed to use SCM as a non-volatile merge cache for the Flash portion of an enterprise SSD. The basic idea is to replace the DRAM memory in the SSD with an SCM and use that SCM to house the FTL Logical Block Map and also as a buffer for writes that are smaller than a Flash page, coalescing writes to the same LBA, and then stage the writes from the SCM to Flash at a later time. Non-volatile merge caching does not require changes to the file system or the block-device interface and can improve performance (due to the lower write latencies offered by the SCM) and reduce the number of erases compared to a Flash-only SSD due to the in-place write property SCMs.

The architecture of the SSD with the SCM-based merge cache is shown in Figure 7.2. The SCM houses the FTL data structures (e.g., the Logical Block Map), the merge cache, and auxiliary bookkeeping structures that are used to manage the merge cache. Each entry in the Logical Block Map on the SCM stores a bit that indicates whether any sectors (1 sector = 512 bytes) within a page reside in the merge cache. An inverted index is used to map a physical sector within the SCM to an LBA. If the inverted index contains an entry for a given LBA, its location in the merge cache is given by the entry's SCM physical sector. The SRAM within the SSD controller is used to cache portions of the inverted index. On an I/O request to the SSD, the FTL is consulted to check whether the requested data is housed in the Flash portion of the SSD or in the merge cache. If the data is in the merge cache, the SRAM cache is probed using the logical page address to find the merge cache

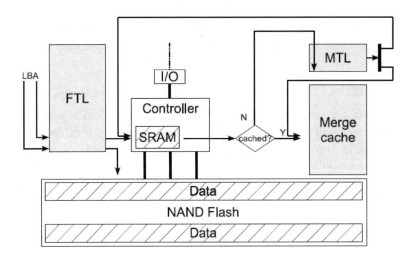

Figure 7.2:   Read/Write Data Access Path for a Flash SSD with a Non-Volatile Merge Cache. Image Source: [191].

entry for that page. On a merge cache miss, a linear search of the inverted index is performed to reconstruct the missing cache entry.

7.4.2   NEW SSD INTERFACES

Traditional storage stacks are designed based on storage device abstractions that are representative of HDDs (rotating disks and moving arms). However, these abstractions are no longer applicable when the underlying storage medium is a non-volatile memory (Flash or SCM). When SSDs were introduced into the market, the goal was to design them such that they provided backward compatibility with their HDD counterparts so that one can use SSDs as drop-in replacements. However, as the share of the overall market has grown for SSDs and with the advent of SCM technologies, there has been growing interest in revisiting the design of the storage stack in light of the new abstraction they provide.

One example of an SSD-specific extension to the drive interface is the `trim` command. The `trim` command improves performance by allowing the operating system to indicate to the SSD the blocks that correspond to files that have been deleted. The SSD can then use this information to optimize the garbage collection process. Prabhakaran et al. [149] explored the use of a transactional interface for Flash-based SSDs. They proposed a cyclic commit protocol that requires less storage overheads than traditional commit protocols and also a file system that can leverage their transactional interface. Given that the FTL decouples the logical block addresses from their physical mapping on the SSD and the management of the stored data is done by the FTL, there have also been proposals to use a semantically richer object-based storage interface for SSDs, rather than block-based interfaces that are more germane to HDDs [2].

While the `trim` command is supported by several SSD vendors, the more radical changes proposed in the research literature will require more extensive changes to the software stack and therefore give rise to backward compatibility issues. Replacing Flash with an SCM (either partly or completely) can have a significant impact on storage interfaces and software. The next section considers a case study of one such prototype SSD - *Onyx* - that uses Phase Change Memory as the underlying storage medium.

7.4.3   CASE STUDY: THE ONYX PCM-BASED SSD

Developing hardware prototypes is an important part of computer architecture and systems research. As non-volatile memory technologies mature from research to production, it is important to study how real chips that use those technologies will behave when used in building a hardware device. One of the first efforts into prototyping and studying a SSD that uses SCM as the storage medium is the *Moneta* project at the University of California, San Diego [28]. Moneta uses DRAM chips to emulate next-generation Phase Change Memory, similar to those discussed in the previous chapters. A key insight from evaluating Moneta is that the operating system, which is designed under the assumption that disks are slow, can become a significant bottleneck to I/O performance when the underlying storage medium is PCM. Their work demonstrated that future SSDs will require a co-design of the hardware and the software, with careful thought to the interface provided by the SSDs and the data structures and algorithms used in the software part of the storage stack.

In their follow-up work, the UCSD researchers developed a prototype SSD, called *Onyx*, which is based on the Moneta architecture and software stack, but with real Micron™ PCM chips [5]. These PCM chips use a NOR Flash style interface. The architecture of Onyx is given in Figure 7.3. Onyx consists of several PCM memory modules, each of which is composed of several 16 MB PCM chips. A pair of memory modules are connected to a PCM memory controller. In addition to managing the PCM chips, each memory controller also implements Start-Gap wear leveling (discussed in Section 4.3). Request scheduling and communication with the memory modules is performed by the Onyx controller (labelled as "Brain" in the Figure).

The performance of Onyx vis-a-vis a Flash-based FusionIO™ ioDrive and Moneta is given in Figure 7.4. While the graphs indicate that Onyx is slower than Moneta, since the PCM chips currently available are slower than those that are projected for the future (which Moneta emulates), Onyx provides better performance than the Flash-based SSD for a wide range of read and write request sizes. The exception is for large write sizes, where the ioDrive provides higher performance. The authors attribute this trend to the greater write bandwidth of ioDrive. The authors also find that the CPU overhead with Onyx is less than with the ioDrive, since the latter requires a CPU-intensive driver whereas Onyx does not.

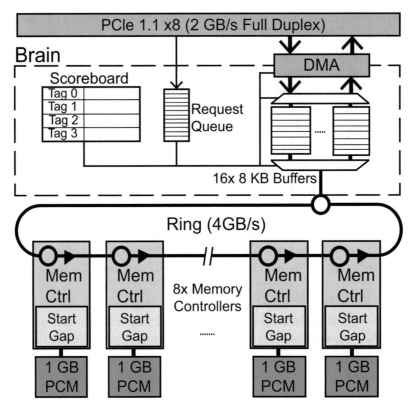

Figure 7.3: Architecture of Onyx. (Image adapted from [5].)

7.4.4 DISCUSSION

As the roadmap for Flash memory scaling approaches its end, we are likely to witness an evolutionary change in SSD architectures from Flash-based devices to those that use SCM. SCMs will continue to provide the energy-efficiency and density benefits that Flash has provided, while also paving the way for improvements in performance and endurance. However, the shift to SCM-based SSDs will also make it increasingly difficult to preserve the legacy block-based storage interfaces and operating systems code that assumes abstractions that are representative of slow electro-mechanical HDDs, as the Moneta and Onyx projects have demonstrated. This is likely to spur further research into storage interfaces and architectures and a change in the storage standards as SCMs gain traction in the storage market.

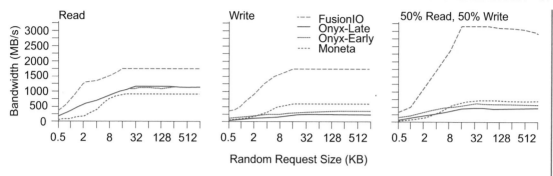

Figure 7.4: Onyx Performance for various read and write request sizes. (Figure adapted from [5].)

7.5   IMPLICATIONS OF MEMORY SYSTEM NON-VOLATILITY ON SYSTEM DESIGN

Rotating magnetic disks, Flash memory, and SCMs share one fundamental property - they are all *non-volatile*. Moreover, in a typical memory hierarchy, all the layers above storage are volatile, since they use SRAM and DRAM for high performance, whereas the storage system (which could itself be a hierarchy consisting of SSDs, rotating disks, and tape) is non-volatile. This abstract view of the memory hierarchy has been the basis for computer architecture and systems software design for many years. For example, in order to put a machine into hibernation to save power, the contents of main memory have to be written to disk before the system enters that state. However, the use of an SCM at the cache or main memory levels of the hierarchy breaks this boundary between the volatile and non-volatile layers, thereby altering the abstraction of the memory hierarchy itself. This fundamental change in the definition of the memory hierarchy can have a significant impact on computer architecture and software systems.

We now discuss three different scenarios where the non-volatility property of SCMs can affect system design, with one case study of each from the literature. The first scenario we discuss deals with file system design when main memory is non-volatile. The second scenario we present relates to how non-volatility can impact the interface that the memory system provides to software. Here we discuss the design of a persistent object store that uses SCM rather than a disk. The final scenario relates to non-volatility as a design knob in hardware, where retention time can be traded off for improvements in performance and power.

7.5.1   FILE SYSTEM DESIGN

File systems are designed based on the assumption that main memory is fast, byte-addressable, and volatile whereas storage is slow, block-addressable, and non-volatile. For example, file systems use main memory as a buffer cache for data read from and to be written to disk but require data to be explicitly flushed to disk in order to consider a write to have committed. The use of an SCM in main explicitly flushed to disk in order to consider a write to have committed. The use of an SCM in main memory blurs these distinctions. Condit et al. [36] demonstrate how one can build a file system for a machine that uses SCM in main memory. Their file system, called *BPFS*, has two design highlights:

- BPFS uses an architecture where both DRAM and an SCM are co-located and are addressable by the CPU, with the physical address space partitioned between the two regions. BPFS places file system data structures that require persistence (i.e., non-volatility) on the SCM whereas DRAM is used for other data.

- Since non-volatility at the main memory level introduces a characteristic that is not present in systems that do not use an SCM, extra care needs to be taken to ensure that writes to SCM, which are persistent, do not leave main memory in an inconsistent state after a power failure or a system crash. For example, in most computer architectures, caches reorder writes to main memory to improve performance. However, the ordering of writes becomes a critical issue when the memory being written to is an SCM. BPFS assumes hardware support for atomicity and ordering. Ordering is provided through *Epoch Barriers*, which allow writes within a barrier to be reordered but the epochs themselves are serialized and written in order to memory. Hardware support for failure atomicity is provided through a capacitor in the SCM memory module. In the event of a power failure, writes to the SCM that are at most 8 bytes in size are guaranteed to complete.

BPFS uses the memory system architecture and the hardware support for atomicity and ordering to implement a variant of shadow paging called *Short Circuit Shadow Paging (SCSP)*. Shadow paging [123] is a technique for facilitating crash recovery and is illustrated in Figure 7.5(a). In shadow paging, every write to a data item results in a copy-on-write. In the case of a file system tree, a write to the file data (which are at the leaves of the tree), will result in a copy-on-write, as the Figure indicates. Then the write to the internal node that points to the leaves will again result in a copy-on-write and so on till the root node of the file system is updated. Therefore, a system crash at any point of time before the creation of the new root node will not affect the state of the file system. However, the key drawback of shadow paging is the fact that every data write will trigger multiple copy-on-writes, from the leaves, all the way up to the root of the tree.

SCSP reduces the overheads of shadow paging by leveraging the fact that failure atomicity is guaranteed by the use of the capacitor on the SCM memory module and therefore writes that are 8 bytes or less can be done *in-place*, instead of requiring a copy-on-write. This is shown in Figure 7.5(b). The file system operations are implemented using epoch barriers to specify the ordering constraints. These ordering constraints are enforced by pinning epochs within the volatile cache hierarchy and serializing their updates to main memory based on their epoch sequence numbers.

7.5.2   THE SOFTWARE INTERFACE TO SCM

In the previous section, we discussed how blurring the distinctions between main memory and storage can affect file system design. More generally, the non-volatility of SCMs coupled with

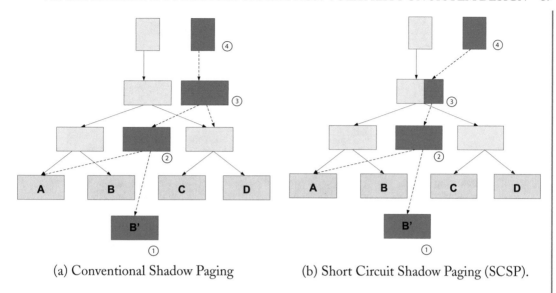

(a) Conventional Shadow Paging   (b) Short Circuit Shadow Paging (SCSP).

Figure 7.5: Shadow Paging in BPFS. The numbers indicate the sequence in which the nodes of the file system tree are updated. The solid arrows correspond to the pointers before the update to data blocks *B* and the dashed arrows are the pointers after the update. (Images adapted from [36] and their conference talk slides.)

their byte-addressability raises questions about the interface that the memory system provides to software. One possibility that several researchers have recently considered is to leverage the properties of SCM to implement a transactional interface, with full ACID (Atomicity, Consistency, Isolation, Durability) semantics [35, 203, 204]. Persistent object stores are useful for several applications, such as search engines, bioinformatics workloads, and scientific computing [176]. Prior work on persistent object stores have attempted to provide the ability to create and manipulate data structures with transactional semantics without the need for a DBMS [176, 178]. Providing transactional semantics facilitates recoverability from system crashes and providing this guarantee without having to use an entire DBMS provides ease of use and reduces performance overheads. These persistent object stores were designed under traditional system assumptions that rotating disks are the only non-volatile storage medium and therefore required movement of data between main memory and disk to provide persistence (non-volatility). Using SCM in the memory system relaxes the need to serialize and deserialize the data between memory and disk and can facilitate the design of persistent data structures using a programming interface that is similar to that used in conventional volatile memory systems for manipulating data structures. We now consider a case study of a specific persistent object store designed for SCMs called *NV-Heaps* [35].

The high-level design of NV-Heaps is shown in Figure 7.6. NV-Heaps implements a persistent object store that programmers can access via an application-level interface. This interface allows

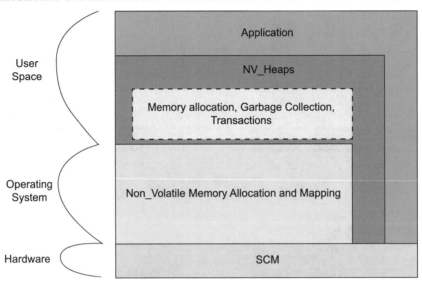

Figure 7.6: Design of NV-Heap based system. (Image redrawn from [35].)

programmers to allocate memory and explicitly designate whether the memory allocated is to be non-volatile (i.e., on SCM) or volatile. While the use of an SCM provides persistence, achieving ACID semantics requires a logging mechanism to undo/redo transactions in the event of a system crash or power failure. NV-Heaps uses a logging mechanism similar to those used in software transactional memory systems [97].

In addition to implementing a persistent object store that provides low-level access to SCM, NV-Heaps also attempts to reduce the likelihood of pointer safety errors by enforcing certain restrictions on the types of pointers that can be created. For example, NV-Heaps disallows having pointers from objects in the non-volatile heap to those in the volatile heap, since those pointers will lead to dangling references when the program completes or across a system reboot. Finally, NV-Heaps uses garbage collection to reclaim the space occupied by dead objects.

In terms of operating system and hardware support, NV-Heaps are treated as ordinary files by the underlying file system. Opening an NV-Heap results in mmap()-ing of the corresponding file into the address space of the application that invokes the call to open the NV-Heap. NV-Heap assumes the availability of epoch barriers (discussed in Section 7.5.2) to provide atomicity and consistency to provide transactional semantics.

7.5.3   NON-VOLATILITY AS A DESIGN KNOB

Having discussed software and programmability issues, we now look at non-volatility from a different angle. The discussion that follows summarizes the work from the literature that exists for a specific

Non-Volatile Memory, namely, STT-RAM. However, the ideas discussed here might be applicable other NVMs too.

Any memory technology can be characterized using several figures-of-merit, including: performance (access times), density (memory cell size), leakage and dynamic power consumption, endurance, and the data retention time. *Data retention time* is the amount of time for which the data in the memory cell is preserved when the external power source is removed and is therefore a measure of non-volatility. SCMs are usually designed for retention times of approximately 10 years. While long retention times are critical when SCMs are used in the storage system, which have stringent long-term data preservation requirements, they are less critical for caches and main memory. Smullen et al. [192] demonstrated that retention time can be traded off for improvements in performance and write energy in STT-RAM to build high-performance, low-power caches for a multicore processor.

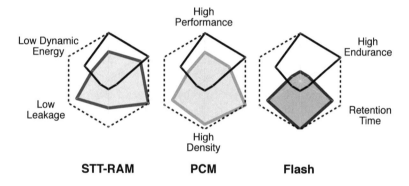

Figure 7.7: A comparison of Non-Volatile Memory Technologies to SRAM. Image Source: [192].

Figure 7.7 compares three non-volatile memories - STT-RAM, PCM, and Flash - to SRAM along several figures of merit. The outer boundary of the spider chart represents the ideal characteristics of an "Universal Memory". The thick black line corresponds to SRAM. The data for this spider chart was obtained from the ITRS Roadmap and other publications on these memory technologies [8]. As this figure indicates, while the non-volatile memories provide significant density and leakage power advantages over SRAM, they do not have the high performance and endurance of SRAM. Among the NVMs, STT-RAM has the highest endurance and is therefore the most suitable for use in the higher layers of the memory system (e.g., caches), since it can withstand the higher write intensities experienced by those layers. However, STT-RAM suffers from high write energies and significantly slower writes than SRAM.

STT-RAM uses Magnetic Tunnel Junctions (MTJs) as their storage element, as explained in Section 1.3. Data is stored in an MTJ as the relative orientation of the magnetization of the pinned and free layers. STT-RAM uses *spin-transfer torque* to change the magnetic orientation of the free layer by passing a large, directional write current through the MTJ. The switching process is regulated by a thermally controlled stochastic process and therefore the free layer could flip at

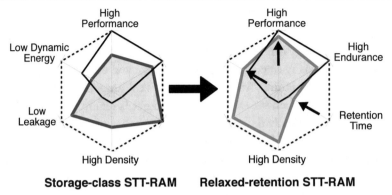

Figure 7.8: Benefits of Reducing the Non-Volatility of STT-RAM. Image Source: [192].

anytime. However, the magnetic properties and device size of the MTJ are chosen to minimize the likelihood of such an event.

The retention time of an MTJ is determined by its *thermal stability*. A high thermal stability means that the memory cell is unlikely to suffer from random bit flips. However, the drawback of a high thermal stability is that a large write current and/or long write times are required to change the state of the MTJ. On the other hand, a lower stability can reduce the write current and latency, but at the cost of lower retention time. This tradeoff is visually depicted in Figure 7.8. A memory array designed with such reduced-retention STT-RAM cells can provide low latency access, which is important for caches, although the short retention time can lead to premature data loss. In order to avoid data loss, Smullen et al. proposed a DRAM-style refresh mechanism which periodically reads all cache lines and writes them back. The refresh frequency is chosen based on the retention time of the cells used for the memory arrays and the capacity of the caches.

Implications for System Design: A key implication of this work is that it is indeed possible to use an SCM for caches, leveraging the leakage power and density benefits they offer, without sacrificing performance or dynamic power. This paves the way for non-volatility at several levels of the memory hierarchy and therefore provides a new abstraction of the memory system. This new abstraction can bring both benefits and challenges to designing systems. For example, the buffering of epochs in volatile caches before releasing them into the SCM main memory in BPFS that we discussed in Section 7.5.1 no longer provides the same guarantees in a machine that has non-volatile caches and therefore warrants a further rethinking of file system design. Moreover, since retention time is traded off to gain performance and energy benefits, non-volatility is no longer a single value (say, 10 years), but is now a variable. Therefore, the *degree of non-volatility* becomes a new metric in defining the configuration of the cache hierarchy. While all layers of the hierarchy will be non-volatile to some extent, those that are closer to the processor will have a lower degree of non-volatility (due to the need for higher performance) whereas those farther away will have higher degrees of non-volatility.

This new abstraction of the memory hierarchy can have interesting implications on the design of software systems.

7.5.4 DISCUSSION

The presence of non-volatility in erstwhile volatile layers of the memory hierarchy is a fundamental design change. We have outlined some of the research on the design of software systems that leverage this property to construct file systems and persistent object stores. We have presented ideas related to using non-volatility as a design knob to architect high-performance and energy efficient memory hierarchies. As several of these works have indicated, effectively leveraging SCMs to at the system level requires careful co-design across the device, architecture, and the software layers.

Bibliography

[1] *International Technology Roadmap for Semiconductors, ITRS 2007.* Cited on page(s) 43

[2] A. Rajimwale and V. Prabhakaran and J.D. Davis. Block Management in Solid-State Drives. In *Proceedings of the USENIX Annual Technical Conferences (USENIX)*, June 2009. Cited on page(s) 85

[3] David Adler, Heinz K. Henisch, and Sir Nevill Mott. The mechanism of threshold switching in amorphous alloys. *Rev. Mod. Phys.*, 50(2):209–220, Apr 1978. DOI: 10.1103/RevModPhys.50.209 Cited on page(s) 19

[4] N. Agrawal and et al. Design Tradeoffs for SSD Performance. In *Proceedings the USENIX Technical Conference (USENIX)*, June 2008. Cited on page(s) 81, 82

[5] A. Akel, A.M. Caulfield, T.I. Mollov, R.K. Gupta, and S. Swanson. Onyx: A Prototype Phase Change Memory Storage Array. In *Workshop on Hot Topics in Storage and File Systems (HotStorage)*, June 2011. Cited on page(s) 79, 85, 86, 87

[6] B. Amir Parviz, D. Ryan, and G.M. Whitesides. Using self-assembly for the fabrication of nano-scale electronic and photonic devices. *Advanced Packaging, IEEE Transactions on*, 26(3):233 – 241, 2003. DOI: 10.1109/TADVP.2003.817971 Cited on page(s) 1

[7] K. Aratani, K. Ohba, T. Mizuguchi, S. Yasuda, T. Shiimoto, T. Tsushima, T. Sone, K. Endo, A. Kouchiyama, S. Sasaki, A. Maesaka, N. Yamada, and H. Narisawa. A novel resistance memory with high scalability and nanosecond switching. In *Electron Devices Meeting, 2007. IEDM 2007. IEEE International*, pages 783 –786, dec. 2007. DOI: 10.1109/IEDM.2007.4419064 Cited on page(s) 16

[8] W. Arden and et al. Semiconductor Industries Association - International Technology Roadmap for Semiconductors, 2009. http://www.itrs.net. Cited on page(s) xi, 91

[9] G. Atwood and R. Bez. 90nm Phase Change Technology with μTrench and Lance Cell Elements. In *VLSI Technology, Systems and Applications, 2007. VLSI-TSA 2007. International Symposium on*, pages 1 –2, 2007. DOI: 10.1109/VTSA.2007.378938 Cited on page(s) 21

[10] Manu Awasthi, Manjunath Shevgoor, Kshitij Sudan, Rajeev Balasubramonian, Bipin Rajendran, and Viji Srinivasan. 'Handling PCM Resistance Drift with Device, Circuit, Architecture, and System Solutions. In *Non-Volatile Memory Workshop, 2011*, 2011. Cited on page(s) 23

[11] M. N. Baibich, J. M. Broto, A. Fert, F. Nguyen Van Dau, F. Petroff, P. Etienne, G. Creuzet, A. Friederich, and J. Chazelas. Giant Magnetoresistance of (001)Fe/(001)Cr Magnetic Super-lattices. *Phys. Rev. Lett.*, 61(21):2472–2475, Nov 1988. DOI: 10.1103/PhysRevLett.61.2472 Cited on page(s) 11

[12] Amir Ban and Ramat Hasharon. Wear leveling of static areas in flash memory. U.S. Patent Number 6,732,221, 2004. Cited on page(s) 44

[13] A. Beck, J. G. Bednorz, Ch. Gerber, C. Rossel, and D. Widmer. Reproducible switching effect in thin oxide films for memory applications. *Applied Physics Letters*, 77(1):139–141, 2000. DOI: 10.1063/1.126902 Cited on page(s) 15

[14] F. Bedeschi, R. Fackenthal, C. Resta, E.M. Donze, M. Jagasivamani, E.C. Buda, F. Pel-lizzer, D.W. Chow, A. Cabrini, G. Calvi, R. Faravelli, A. Fantini, G. Torelli, D. Mills, R. Gastaldi, and G. Casagrande. A bipolar-selected phase change memory featuring multi-level cell storage. *Solid-State Circuits, IEEE Journal of*, 44(1):217 –227, 2009. DOI: 10.1109/JSSC.2008.2006439 Cited on page(s) 22

[15] R. S. Beech, J. A. Anderson, A. V. Pohm, and J. M. Daughton. Curie point written magneto-resistive memory. *Journal of Applied Physics*, 87(9):6403–6405, 2000. DOI: 10.1063/1.372720 Cited on page(s) 13

[16] Avraham Ben-Aroya and Sivan Toledo. Competitive analysis of flash-memory algorithms. In *ESA'06: Proceedings of the 14th conference on Annual European Symposium*, pages 100–111, 2006. DOI: 10.1145/1921659.1921669 Cited on page(s) 44

[17] L. Berger. Emission of spin waves by a magnetic multilayer traversed by a current. *Phys. Rev. B*, 54(13):9353–9358, Oct 1996. DOI: 10.1103/PhysRevB.54.9353 Cited on page(s) 13

[18] A. Bette, J. DeBrosse, D. Gogl, H. Hoenigschmid, R. Robertazzi, C. Arndt, D. Braun, D. Casarotto, R. Havreluk, S. Lammers, W. Obermaier, W. Reohr, H. Viehmann, W.J. Gal-lagher, and G. Muller. A high-speed 128 Kbit MRAM core for future universal memory applications. In *VLSI Circuits, 2003. Digest of Technical Papers. 2003 Symposium on*, pages 217 – 220, 2003. DOI: 10.1109/VLSIC.2003.1221207 Cited on page(s) 12, 13

[19] R. Bez, E. Camerlenghi, A. Modelli, and A. Visconti. Introduction to flash memory. *Pro-ceedings of the IEEE*, 91(4):489 – 502, 2003. DOI: 10.1109/JPROC.2003.811702 Cited on page(s) 6

[20] Andrew H. Bobeck. New concept in large size memory arrays - the twistor. *Journal of Applied Physics*, 29(3):485 –486, March 1958. DOI: 10.1063/1.1723190 Cited on page(s) 11

[21] Simona Boboila and Peter Desnoyers. Write endurance in flash drives: measurements and analysis. In *Proceedings of the 8th USENIX conference on File and storage technologies*, FAST'10, pages 9–9, Berkeley, CA, USA, 2010. USENIX Association. Cited on page(s) 4, 8

[22] M. Breitwisch, T. Nirschl, C.F. Chen, Y. Zhu, M.H. Lee, M. Lamorey, G.W. Burr, E. Joseph, A. Schrott, J.B. Philipp, R. Cheek, T.D. Happ, S.H. Chen, S. Zaidr, P. Flaitz, J. Bruley, R. Dasaka, B. Rajendran, S. Rossnage, M. Yang, Y.C. Chen, R. Bergmann, H.L. Lung, and C. Lam. Novel Lithography-Independent Pore Phase Change Memory. In *VLSI Technology, 2007 IEEE Symposium on*, pages 100–101, 2007. DOI: 10.1109/VLSIT.2007.4339743 Cited on page(s) 21

[23] J.E. Brewer and M. Gill, editors. *Nonvolatile Memory Technologies with Emphasis on Flash*. IEEE Press, 2008. Cited on page(s) 4, 6, 9

[24] Dudley A. Buck. Ferroelectrics for Digital Information Storage and Switching. M. Thesis in Electrical Engineering, M.I.T, June 1952. Cited on page(s) 9

[25] G. W. Burr, B. N. Kurdi, J. C. Scott, C. H. Lam, K. Gopalakrishnan, and R. S. Shenoy. Overview of candidate device technologies for storage-class memory. *IBM Journal of Research and Development*, 52(4.5):449–464, 2008. DOI: 10.1147/rd.524.0449 Cited on page(s) 9

[26] Geoffrey W. Burr, Matthew J. Breitwisch, Michele Franceschini, Davide Garetto, Kailash Gopalakrishnan, Bryan Jackson, Blent Kurdi, Chung Lam, Luis A. Lastras, Alvaro Padilla, Bipin Rajendran, Simone Raoux, and Rohit S. Shenoy. Phase Change Memory Technology. *J. Vac. Sci. Technol. B*, 28, 2010. DOI: 10.1116/1.3301579 Cited on page(s) 1, 3, 17, 18

[27] C. Dirik and B. Jacob. The Performance of PC Solid-State Disks (SSDs) as a Function of Bandwidth, Concurrency, Device Architecture, and System Organization. In *Proceedings of the International Symposium on Computer Architecture (ISCA)*, pages 279–289, June 2009. DOI: 10.1145/1555815.1555790 Cited on page(s) 81

[28] A.M. Caulfield, A. De, J. Coburn, T.I. Mollov, R.K. Gupta, and S. Swanson. Moneta: A High-Performance Storage Array Architecture for Next-Generation, Non-Volatile Memories. In *Proceedings of the International Symposium on Microarchitecture (MICRO)*, pages 385–395, December 2010. DOI: 10.1109/MICRO.2010.33 Cited on page(s) 85

[29] Y.C. Chen, C.T. Rettner, S. Raoux, G.W. Burr, S.H. Chen, R.M. Shelby, M. Salinga, W.P. Risk, T.D. Happ, G.M. McClelland, M. Breitwisch, A. Schrott, J.B. Philipp, M.H. Lee, R. Cheek, T. Nirschl, M. Lamorey, C.F. Chen, E. Joseph, S. Zaidi, B. Yee, H.L. Lung, R. Bergmann, and C. Lam. Ultra-Thin Phase-Change Bridge Memory Device Using GeSb. In *Electron Devices Meeting, 2006. IEDM '06. International*, pages 1–4, 2006. DOI: 10.1109/IEDM.2006.346910 Cited on page(s) 19, 21

[30] C.J. Chevallier, Chang Hua Siau, S.F. Lim, S.R. Namala, M. Matsuoka, B.L. Bateman, and D. Rinerson. A 0.13 μm 64Mb multi-layered conductive metal-oxide memory. In *Solid-State Circuits Conference Digest of Technical Papers (ISSCC), 2010 IEEE International*, pages 260–261, 2010. DOI: 10.1109/ISSCC.2010.5433945 Cited on page(s) 14, 15, 16

[31] A. Chimenton, C. Zambelli, P. Olivo, and A. Pirovano. Set of electrical characteristic parameters suitable for reliability analysis of multimegabit phase change memory arrays. In *Non-Volatile Semiconductor Memory Workshop, 2008 and 2008 International Conference on Memory Technology and Design. NVSMW/ICMTD 2008. Joint*, pages 49 –51, May 2008. DOI: 10.1109/NVSMW.2008.20 Cited on page(s) 23

[32] Stephen Y. Chou, Peter R. Krauss, and Preston J. Renstrom. Imprint Lithography with 25-Nanometer Resolution. *Science*, 272(5258):85–87, 1996. DOI: 10.1126/science.272.5258.85 Cited on page(s) 1

[33] Suock Chung, K.-M. Rho, S.-D. Kim, H.-J. Suh, D.-J. Kim, H.-J. Kim, S.-H. Lee, J.-H. Park, H.-M. Hwang, S.-M. Hwang, J.-Y. Lee, Y.-B. An, J.-U. Yi, Y.-H. Seo, D.-H. Jung, M.-S. Lee, S.-H. Cho, J.-N. Kim, G.-J. Park, Gyuan Jin, A. Driskill-Smith, V. Nikitin, A. Ong, X. Tang, Yongki Kim, J.-S. Rho, S.-K. Park, S.-W. Chung, J.-G. Jeong, and S.-J. Hong. Fully integrated 54nm STT-RAM with the smallest bit cell dimension for high density memory application. In *Electron Devices Meeting (IEDM), 2010 IEEE International*, pages 12.7.1 –12.7.4, 2010. DOI: 10.1109/IEDM.2010.5703351 Cited on page(s) 13, 16

[34] G.F. Close, U. Frey, M. Breitwisch, H.L. Lung, C. Lam, C. Hagleitner, and E. Eleftheriou. Device, circuit and system-level analysis of noise in multi-bit phase-change memory. In *Electron Devices Meeting (IEDM), 2010 IEEE International*, pages 29.5.1 –29.5.4, 2010. DOI: 10.1109/IEDM.2010.5703445 Cited on page(s) 24

[35] J. Coburn, A.M. Caulfield, A. Akel, L.M. Grupp, R.K. Gupta, R. Jhala, and S. Swanson. NV-Heaps: Making Persistent Objects Fast and Safe with Next-Generation, Non-Volatile Memories. In *Proceedings of the International Conference on Architectural Support for Programming Languages and Operating Systems (ASPLOS)*, pages 105–118, March 2011. DOI: 10.1145/1961296.1950380 Cited on page(s) 89, 90

[36] J. Condit, E.B. Nightingale, C. Frost, E. Ipek, D. Burger, B. Lee, and D. Coetzee. Better I/O Through Byte-Addressable, Persistent Memory. In *Proceedings of the Symposium on Operating Systems Principles (SOSP)*, pages 133–146, October 2009. DOI: 10.1145/1629575.1629589 Cited on page(s) 88, 89

[37] M. Dawber, K. M. Rabe, and J. F. Scott. Physics of thin-film ferroelectric oxides. *Rev. Mod. Phys.*, 77(4):1083–1130, Oct 2005. DOI: 10.1103/RevModPhys.77.1083 Cited on page(s) 11

[38] I.H. Doh, J. Choi, D. Lee, and S.H. Noh. Exploiting Non-Volatile RAM to Enhance Flash File System Performance. In *Proceedings of the International Conference on Embedded Software (EMSOFT)*, pages 164–173, October 2007. DOI: 10.1145/1289927.1289955 Cited on page(s) 83

[39] M. Durlam, D. Addie, J. Akerman, B. Butcher, P. Brown, J. Chan, M. DeHerrera, B.N. Engel, B. Feil, G. Grynkewich, J. Janesky, M. Johnson, K. Kyler, J. Molla, J. Martin, K. Nagel, J. Ren, N.D. Rizzo, T. Rodriguez, L. Savtchenko, J. Salter, J.M. Slaughter, K. Smith, J.J. Sun, M. Lien, K. Papworth, P. Shah, W. Qin, R. Williams, L. Wise, and S. Tehrani. A 0.18 μm 4Mb toggling MRAM. In *Electron Devices Meeting, 2003. IEDM '03 Technical Digest. IEEE International*, pages 34.6.1 – 34.6.3, 2003. DOI: 10.1109/IEDM.2003.1269448 Cited on page(s) 12

[40] M. Durlam, B. Craigo, M. DeHerrera, B.N. Engel, G. Grynkewich, B. Huang, J. Janesky, M. Martin, B. Martino, J. Salter, J.M. Slaughter, L. Wise, and S. Tehrani. Toggle mram: A highly-reliable non-volatile memory. In *VLSI Technology, Systems and Applications, 2007. VLSI-TSA 2007. International Symposium on*, pages 1 –2, 2007. DOI: 10.1109/VTSA.2007.378942 Cited on page(s) 12

[41] T. Eshita, K. Nakamura, M. Mushiga, A. Itho, S. Miyagaki, H. Yamawaki, M. Aoki, S. Kishii, and Y. Arimoto. Fully functional 0.5- μm 64-kbit embedded SBT FeRAM using a new low temperature SBT deposition technique. In *VLSI Technology, 1999. Digest of Technical Papers. 1999 Symposium on*, pages 139 –140, 1999. DOI: 10.1109/VLSIT.1999.799382 Cited on page(s) 11

[42] P. Fantini, G. Betti Beneventi, A. Calderoni, L. Larcher, P. Pavan, and F. Pellizzer. Characterization and modelling of low-frequency noise in PCM devices. In *Electron Devices Meeting, 2008. IEDM 2008. IEEE International*, pages 1 –4, 2008. DOI: 10.1109/IEDM.2008.4796656 Cited on page(s) 24

[43] G. A. Fedde. Design of a 1.5 Million Bit Plated Wire Memory. *Journal of Applied Physics*, 37(3):1373 –1375, March 1966. DOI: 10.1063/1.1708477 Cited on page(s) 11

[44] I. Friedrich, V. Weidenhof, W. Njoroge, P. Franz, and M. Wuttig. Structural transformations of $Ge_2Sb_2Te_5$ films studied by electrical resistance measurements. *Journal of Applied Physics*, 87(9):4130–4134, 2000. DOI: 10.1063/1.373041 Cited on page(s) 18

[45] E. Gal and S. Toledo. Algorithms and Data Structures for Flash Memories. *ACM Computing Surveys*, 37(2):138–163, June 2005. DOI: 10.1145/1089733.1089735 Cited on page(s) 44, 82

[46] W. J. Gallagher and S. S. P. Parkin. Development of the magnetic tunnel junction MRAM at IBM: From first junctions to a 16-Mb MRAM demonstrator chip. *IBM Journal of Research and Development*, 50(1):5 – 23, 2006. DOI: 10.1147/rd.501.0005 Cited on page(s) 12

[47] B. Gao, W.Y. Chang, B. Sun, H.W. Zhang, L.F. Liu, X.Y. Liu, R.Q. Han, T.B. Wu, and J.F. Kang. Identification and application of current compliance failure phenomenon in RRAM

device. In *VLSI Technology Systems and Applications (VLSI-TSA), 2010 International Symposium on*, pages 144 –145, 2010. DOI: 10.1109/VTSA.2010.5488912 Cited on page(s) 14

[48] J. F. Gibbons and W. E. Beadle. Switching properties of thin NiO films. *Solid-State Electronics*, 7(11):785 – 790, 1964. DOI: 10.1016/0038-1101(64)90131-5 Cited on page(s) 14

[49] P. Gonon, M. Mougenot, C. Vallée, C. Jorel, V. Jousseaume, H. Grampeix, and F. El Kamel. Resistance switching in HfO_2 metal-insulator-metal devices. *Journal of Applied Physics*, 107(7):074507, 2010. DOI: 10.1063/1.3357283 Cited on page(s) 14

[50] K. Gopalakrishnan, R.S. Shenoy, C.T. Rettner, R.S. King, Y. Zhang, B. Kurdi, L.D. Bozano, J.J. Welser, M.E. Rothwell, M. Jurich, M.I. Sanchez, M. Hernandez, P.M. Rice, W.P. Risk, and H.K. Wickramasinghe. The micro to nano addressing block (MNAB). In *Electron Devices Meeting, 2005. IEDM Technical Digest. IEEE International*, pages 471 –474, 2005. DOI: 10.1109/IEDM.2005.1609382 Cited on page(s) 4

[51] K. Gopalakrishnan, R.S. Shenoy, C.T. Rettner, K. Virwani, D.S. Bethune, R.M. Shelby, G.W. Burr, A. Kellock, R.S. King, K. Nguyen, A.N. Bowers, M. Jurich, B. Jackson, A.M. Friz, T. Topuria, P.M. Rice, and B.N. Kurdi. Highly-scalable novel access device based on Mixed Ionic Electronic conduction (MIEC) materials for high density phase change memory (PCM) arrays. In *VLSI Technology (VLSIT), 2010 Symposium on*, pages 205 –206, 2010. DOI: 10.1109/VLSIT.2010.5556229 Cited on page(s) 4, 19

[52] L. Goux, D. Tio Castro, G. Hurkx, J.G. Lisoni, R. Delhougne, D.J. Gravesteijn, K. Attenborough, and D.J. Wouters. Degradation of the reset switching during endurance testing of a phase-change line cell. *Electron Devices, IEEE Transactions on*, 56(2):354 –358, 2009. DOI: 10.1109/TED.2008.2010568 Cited on page(s) 20, 24

[53] D. Grice, H. Brandt, C. Wright, P. McCarthy, A. Emerich, T. Schimke, C. Archer, J. Carey, P. Sanders, J. A. Fritzjunker, S. Lewis, and P. Germann. Breaking the petaflops barrier. *IBM Journal of Research and Development*, 53(5):1:1 –1:16, 2009. DOI: 10.1147/JRD.2009.5429067 Cited on page(s) 1

[54] Musarrat Hasan, Rui Dong, H. J. Choi, D. S. Lee, D.-J. Seong, M. B. Pyun, and Hyunsang Hwang. Uniform resistive switching with a thin reactive metal interface layer in metal-$La_{0.7}Ca_{0.3}MnO_3$-metal heterostructures. *Applied Physics Letters*, 92(20):202102, 2008. DOI: 10.1063/1.2932148 Cited on page(s) 14

[55] T. Hayashi, Y. Igarashi, D. Inomata, T. Ichimori, T. Mitsuhashi, K. Ashikaga, T. Ito, M. Yoshimaru, M. Nagata, S. Mitarai, H. Godaiin, T. Nagahama, C. Isobe, H. Moriya, M. Shoji, Y. Ito, H. Kuroda, and M. Sasaki. A novel stack capacitor cell for high density FeRAM compatible with CMOS logic. In *Electron Devices Meeting, 2002. IEDM '02. Digest. International*, pages 543 – 546, 2002. DOI: 10.1109/IEDM.2002.1175899 Cited on page(s) 9

[56] Y. Hosoi, Y. Tamai, T. Ohnishi, K. Ishihara, T. Shibuya, Y. Inoue, S. Yamazaki, T. Nakano, S. Ohnishi, N. Awaya, H. Inoue, H. Shima, H. Akinaga, H. Takagi, H. Akoh, and Y. Tokura. High Speed Unipolar Switching Resistance RAM (RRAM) Technology. In *Electron Devices Meeting, 2006. IEDM '06. International*, pages 1 –4, 2006. DOI: 10.1109/IEDM.2006.346732 Cited on page(s) 14

[57] M. Hosomi, H. Yamagishi, T. Yamamoto, K. Bessho, Y. Higo, K. Yamane, H. Yamada, M. Shoji, H. Hachino, C. Fukumoto, H. Nagao, and H. Kano. A novel nonvolatile memory with spin torque transfer magnetization switching: spin-ram. In *Electron Devices Meeting, 2005. IEDM Technical Digest. IEEE International*, pages 459 –462, 2005. DOI: 10.1109/IEDM.2005.1609379 Cited on page(s) 13

[58] Y. Hu and M.H. White. A new buried-channel EEPROM device. *Electron Devices, IEEE Transactions on*, 39(11):2670, November 1992. DOI: 10.1109/16.163541 Cited on page(s) 9

[59] Byungjoon Hwang, Jeehoon Han, Myeong-Cheol Kim, Sunggon Jung, Namsu Lim, Sowi Jin, Yongsik Yim, Donghwa Kwak, Jaekwan Park, Jungdal Choi, and Kinam Kim. Comparison of double patterning technologies in NAND flash memory with sub-30nm node. In *Solid State Device Research Conference, 2009. ESSDERC '09. Proceedings of the European*, pages 269 –271, 2009. DOI: 10.1109/ESSDERC.2009.5331401 Cited on page(s) 9

[60] Y.N. Hwang, C.Y. Um, J.H. Lee, C.G. Wei, H.R. Oh, G.T. Jeong, H.S. Jeong, C.H. Kim, and C.H. Chung. MLC PRAM with SLC write-speed and robust read scheme. In *VLSI Technology (VLSIT), 2010 Symposium on*, pages 201 –202, 2010. Cited on page(s) 22

[61] D. Ielmini, S. Lavizzari, D. Sharma, and A.L. Lacaita. Physical interpretation, modeling and impact on phase change memory (PCM) reliability of resistance drift due to chalcogenide structural relaxation. In *Electron Devices Meeting, 2007. IEDM 2007. IEEE International*, pages 939 –942, 2007. DOI: 10.1109/IEDM.2007.4419107 Cited on page(s) 23

[62] K. Iida, M. Saito, and K. Furukawa. An 8 MBYTE magnetic bubble memory. *Magnetics, IEEE Transactions on*, 15(6):1892 – 1894, November 1979. DOI: 10.1109/TMAG.1979.1060533 Cited on page(s) 11

[63] S. Ikeda, J. Hayakawa, Y. Ashizawa, Y. M. Lee, K. Miura, H. Hasegawa, M. Tsunoda, F. Matsukura, and H. Ohno. Tunnel magnetoresistance of 604% at 300 K by suppression of Ta diffusion in CoFeB/MgO/CoFeB pseudo-spin-valves annealed at high temperature. *Applied Physics Letters*, 93(8):082508, 2008. DOI: 10.1063/1.2976435 Cited on page(s) 12

[64] D.H. Im, J.I. Lee, S.L. Cho, H.G. An, D.H. Kim, I.S. Kim, H. Park, D.H. Ahn, H. Horii, S.O. Park, U-In Chung, and J.T. Moon. A unified 7.5nm dash-type confined cell for high performance PRAM device. In *Electron Devices Meeting, 2008. IEDM 2008. IEEE International*, pages 1 –4, 2008. DOI: 10.1109/IEDM.2008.4796654 Cited on page(s) 21

102 BIBLIOGRAPHY

[65] Masatoshi Imada, Atsushi Fujimori, and Yoshinori Tokura. Metal-insulator transitions. *Rev. Mod. Phys.*, 70(4):1039–1263, Oct 1998. DOI: 10.1103/RevModPhys.70.1039   Cited on page(s) 15

[66] Engin Ipek, Jeremy Condit, Edmund B. Nightingale, Doug Burger, and Thomas Moscibroda. Dynamically replicated memory: building reliable systems from nanoscale resistive memories. In *ASPLOS-XV*, pages 3–14, 2010. DOI: 10.1145/1735970.1736023   Cited on page(s) 69, 70, 71, 72

[67] T. Ishigaki, T. Kawahara, R. Takemura, K. Ono, K. Ito, H. Matsuoka, and H. Ohno.   A multi-level-cell spin-transfer torque memory with series-stacked magnetotunnel junctions. In *VLSI Technology (VLSIT), 2010 Symposium on*, pages 47 –48, 2010. DOI: 10.1109/VLSIT.2010.5556126   Cited on page(s) 14

[68] H. Ishiwara. Recent progress in ferroelectirc memory technology. In *Solid-State and Integrated Circuit Technology, 2006. ICSICT '06. 8th International Conference on*, pages 713 –716, 2006. DOI: 10.1109/ICSICT.2006.306466   Cited on page(s) 10

[69] ITRS. International Technology Roadmap for Semiconductors. http://www.itrs.net/, June 2009. Cited on page(s) 7

[70] ITRS. Report of Emerging Research Memory Technologies Workshop, 2010. http://www.itrs.net/Links/2010ITRS/2010Update/ToPost/ERD_ERM_ 2010FINALReportMemoryAssessment_ITRS.pdf, June 2010.   Cited on page(s) 14

[71] M. H. Jang, S. J. Park, D. H. Lim, M.-H. Cho, Y. K. Kim, H.-J. Yi, and H. S. Kim. Structural Stability and Phase-Change Characteristics of $Ge_2Sb_2Te_5/SiO_2$ Nano-Multilayered Films. *Electrochemical and Solid-State Letters*, 12(4):H151–H154, 2009. DOI: 10.1149/1.3079479 Cited on page(s) 18

[72] Ronald N. Kalla, Balaram Sinharoy, William J. Starke, and Michael S. Floyd. Power7: Ibm's next-generation server processor. *IEEE Micro*, 30(2):7–15, 2010. DOI: 10.1109/MM.2010.38   Cited on page(s) 61

[73] D-H. Kang, J.-H. Lee, J.H. Kong, D. Ha, J. Yu, C.Y. Um, J.H. Park, F. Yeung, J.H. Kim, W.I. Park, Y.J. Jeon, M.K. Lee, Y.J. Song, J.H. Oh, G.T. Jeong, and H.S. Jeong. Two-bit cell operation in diode-switch phase change memory cells with 90nm technology. In *VLSI Technology, 2008 Symposium on*, pages 98 –99, 2008. DOI: 10.1109/VLSIT.2008.4588577 Cited on page(s) 22

[74] S. Kang, W. Y. Cho, B.-H. Cho, K.-J. Lee, C.-S. Lee, H.-R. Oh, B.-G. Choi, Q. Wang, H.-J. Kim, M.-H. Park, Y. H. Ro, S. Kim, C.-D. Ha, K.-S. Kim, Y.-R. Kim, D.-E. Kim,

C.-K. Kwak, H.-G. Byun, G. Jeong, H. Jeong, K. Kim, and Y. Shin. A 0.1 μm 1.8V 256-Mb Phase-Change Random Access Memory (PRAM) With 66-MHz Synchronous Burst-Read Operation. *Solid-State Circuits, IEEE Journal of*, 42(1):210 –218, 2007. DOI: 10.1109/JSSC.2006.888349 Cited on page(s) 22

[75] S. F. Karg, G. I. Meijer, J. G. Bednorz, C. T. Rettner, A. G. Schrott, E. A. Joseph, C. H. Lam, M. Janousch, U. Staub, F. La Mattina, S. F. Alvarado, D. Widmer, R. Stutz, U. Drechsler, and D. Caimi. Transition-metal-oxide-based resistance-change memories. *IBM Journal of Research and Development*, 52(4.5):481 –492, 2008. DOI: 10.1147/rd.524.0481 Cited on page(s) 14

[76] Y. Kato, T. Yamada, and Y. Shimada. 0.18- μm nondestructive readout FeRAM using charge compensation technique. *Electron Devices, IEEE Transactions on*, 52(12):2616 – 2621, 2005. DOI: 10.1109/TED.2005.859688 Cited on page(s) 10

[77] R. Katsumata, M. Kito, Y. Fukuzumi, M. Kido, H. Tanaka, Y. Komori, M. Ishiduki, J. Matsunami, T. Fujiwara, Y. Nagata, Li Zhang, Y. Iwata, R. Kirisawa, H. Aochi, and A. Nitayama. Pipe-shaped BiCS flash memory with 16 stacked layers and multi-level-cell operation for ultra high density storage devices. In *VLSI Technology, 2009 Symposium on*, pages 136 –137, 2009. Cited on page(s) 9

[78] DerChang Kau, S. Tang, I.V. Karpov, R. Dodge, B. Klehn, J.A. Kalb, J. Strand, A. Diaz, N. Leung, J. Wu, S. Lee, T. Langtry, Kuo wei Chang, C. Papagianni, Jinwook Lee, J. Hirst, S. Erra, E. Flores, N. Righos, H. Castro, and G. Spadini. A stackable cross point Phase Change Memory. In *Electron Devices Meeting (IEDM), 2009 IEEE International*, pages 1 –4, 2009. DOI: 10.1109/IEDM.2009.5424263 Cited on page(s) 25

[79] H. Kawasaki, M. Khater, M. Guillorn, N. Fuller, J. Chang, S. Kanakasabapathy, L. Chang, R. Muralidhar, K. Babich, Q. Yang, J. Ott, D. Klaus, E. Kratschmer, E. Sikorski, R. Miller, R. Viswanathan, Y. Zhang, J. Silverman, Q. Ouyang, A. Yagishita, M. Takayanagi, W. Haensch, and K. Ishimaru. Demonstration of highly scaled FinFET SRAM cells with high-k metal gate and investigation of characteristic variability for the 32 nm node and beyond. In *Electron Devices Meeting, 2008. IEDM 2008. IEEE International*, pages 1 –4, December 2008. DOI: 10.1109/IEDM.2008.4796661 Cited on page(s) 3

[80] Lim Kevin et al. Disaggregated memory for expansion and sharing in blade servers. In *ISCA-36*, pages 267–278, 2009. DOI: 10.1145/1555815.1555789 Cited on page(s) 27

[81] Taeho Kgil, David Roberts, and Trevor Mudge. Improving nand flash based disk caches. In *ISCA '08: Proceedings of the 35th annual international symposium on Computer architecture*, pages 327–338, 2008. DOI: 10.1145/1394608.1382149 Cited on page(s) 44

[82] D. C. Kim, S. Seo, S. E. Ahn, D.-S. Suh, M. J. Lee, B.-H. Park, I. K. Yoo, I. G. Baek, H.-J. Kim, E. K. Yim, J. E. Lee, S. O. Park, H. S. Kim, U-In Chung, J. T. Moon, and B. I. Ryu. Electrical observations of filamentary conductions for the resistive memory switching in NiO films. *Applied Physics Letters*, 88(20):202102, 2006. DOI: 10.1063/1.2204649 Cited on page(s) 14

[83] D. S. Kim, Y. H. Kim, C. E. Lee, and Y. T. Kim. Colossal electroresistance mechanism in a Au/Pr$_{0.7}$Ca$_{0.3}$MnO$_3$/Pt sandwich structure: Evidence for a Mott transition. *Phys. Rev. B*, 74(17):174430, Nov 2006. DOI: 10.1103/PhysRevB.74.174430 Cited on page(s) 15

[84] J. Kim, H.G. Lee, S. Choi, and K.I. Bahng. A PRAM and NAND Flash Hybrid Architecture for High-Performance Embedded Storage Subsystems. In *Proceedings of the International Conference on Embedded Software (EMSOFT)*, pages 31–40, October 2008. DOI: 10.1145/1450058.1450064 Cited on page(s) 83

[85] Kinam Kim. Technology for sub-50nm DRAM and NAND flash manufacturing. In *Electron Devices Meeting, 2005. IEDM Technical Digest. IEEE International*, pages 323 –326, 2005. DOI: 10.1109/IEDM.2005.1609340 Cited on page(s) 9

[86] Kinam Kim and Su Jin Ahn. Reliability investigations for manufacturable high density PRAM. In *Reliability Physics Symposium, 2005. Proceedings. 43rd Annual. 2005 IEEE International*, pages 157 – 162, 17-21 2005. DOI: 10.1109/RELPHY.2005.1493077 Cited on page(s) 23, 24

[87] Kinam Kim and Jungdal Choi. Future Outlook of NAND Flash Technology for 40nm Node and Beyond. In *Non-Volatile Semiconductor Memory Workshop, 2006. IEEE NVSMW 2006. 21st*, pages 9 –11, 2006. DOI: 10.1109/.2006.1629474 Cited on page(s) 1

[88] S. Kim and H.-S.P. Wong. Analysis of temperature in phase change memory scaling. *Electron Device Letters, IEEE*, 28(8):697 –699, 2007. DOI: 10.1109/LED.2007.901347 Cited on page(s) 21

[89] W.I. Kinney, W. Shepherd, W. Miller, J. Evans, and R. Womack. A non-volatile memory cell based on ferroelectric storage capacitors. In *Electron Devices Meeting, 1987 International*, volume 33, pages 850 – 851, 1987. DOI: 10.1109/IEDM.1987.191567 Cited on page(s) 9, 10

[90] M.S. Klamkin and D. J. Newman. Extensions of the birthday surprise. *Journal of Combinatorial Theory*, 1967. DOI: 10.1016/S0021-9800(67)80075-9 Cited on page(s) 56

[91] M.N. Kozicki, Mira Park, and M. Mitkova. Nanoscale memory elements based on solid-state electrolytes. *Nanotechnology, IEEE Transactions on*, 4(3):331 – 338, May 2005. DOI: 10.1109/TNANO.2005.846936 Cited on page(s) 14, 15

[92] Daniel Krebs, Simone Raoux, Charles T. Rettner, Geoffrey W. Burr, Martin Salinga, and Matthias Wuttig. Threshold field of phase change memory materials measured using phase change bridge devices. *Applied Physics Letters*, 95(8):082101, 2009. DOI: 10.1063/1.3210792 Cited on page(s) 19

[93] A.L. Lacaita and D. Ielmini. Reliability issues and scaling projections for phase change non volatile memories. In *Electron Devices Meeting, 2007. IEDM 2007. IEEE International*, pages 157 –160, 2007. Cited on page(s) 23

[94] S. Lai and T. Lowrey. OUM - A 180 nm nonvolatile memory cell element technology for stand alone and embedded applications. In *Electron Devices Meeting, 2001. IEDM Technical Digest. International*, pages 36.5.1 –36.5.4, 2001. DOI: 10.1109/IEDM.2001.979636 Cited on page(s) 1, 17, 24

[95] S. K. Lai. Flash memories: Successes and challenges. *IBM Journal of Research and Development*, 52(4.5):529 –535, 2008. DOI: 10.1147/rd.524.0529 Cited on page(s) 6, 7, 8

[96] D R Lamb and P C Rundle. A non-filamentary switching action in thermally grown silicon dioxide films. *British Journal of Applied Physics*, 18(1):29, 1967. DOI: 10.1088/0508-3443/18/1/306 Cited on page(s) 14

[97] J.R. Larus and R. Rajwar. *Transactional Memory*. Morgan & Claypool Publishers, 2007. Cited on page(s) 90

[98] L.A. Lastras-Montaño, A. Jagmohan, and M.M. Franceschini. An area and latency assessment for coding for memories with stuck cells. In *GLOBECOM Workshops (GC Wkshps), 2010 IEEE*, pages 1851 –1855, 2010. DOI: 10.1109/GLOCOMW.2010.5700262 Cited on page(s) 24

[99] S. Lavizzari, D. Sharma, and D. Ielmini. Threshold-switching delay controlled by current fluctuations in phase-change memory devices. *Electron Devices, IEEE Transactions on*, 57(5):1047 –1054, May 2010. DOI: 10.1109/TED.2010.2042768 Cited on page(s) 24

[100] Benjamin Lee et al. Architecting Phase Change Memory as a Scalable DRAM Alternative. In *ISCA-36*, 2009. DOI: 10.1145/1555815.1555758 Cited on page(s) 29

[101] Chang Hyun Lee, Kyung In Choi, Myoung Kwan Cho, Yun Heub Song, Kyu Charn Park, and Kinam Kim. A novel SONOS structure of $SiO_2/SiN/Al_2O_3$ with TaN metal gate for multi-giga bit flash memories. In *Electron Devices Meeting, 2003. IEDM '03 Technical Digest. IEEE International*, pages 26.5.1 – 26.5.4, 2003. Cited on page(s) 9

[102] Choong-Ho Lee, Suk-Kang Sung, Donghoon Jang, Sehoon Lee, Seungwook Choi, Jonghyuk Kim, Sejun Park, Minsung Song, Hyun-Chul Baek, Eungjin Ahn, Jinhyun Shin, Kwangshik Shin, Kyunghoon Min, Sung-Soon Cho, Chang-Jin Kang, Jungdal Choi, Keonsoo Kim, Jeong-Hyuk Choi, Kang-Deog Suh, and Tae-Sung Jung. A highly manufacturable integration

technology for 27nm 2 and 3bit/cell nand flash memory. In *Electron Devices Meeting (IEDM), 2010 IEEE International*, pages 5.1.1 –5.1.4, dec. 2010. DOI: 10.1109/IEDM.2010.5703299 Cited on page(s) 16

[103] Dongsoo Lee, Dae-Kue Hwang, Man Chang, Yunik Son, Dong jun Seong, Dooho Choi, and Hyunsang Hwang. Resistance switching of Al doped ZnO for Non Volatile Memory applications. In *Non-Volatile Semiconductor Memory Workshop, 2006. IEEE NVSMW 2006. 21st*, pages 86 –87, 2006. DOI: 10.1109/.2006.1629506 Cited on page(s) 15

[104] H.Y. Lee, P.S. Chen, T.Y. Wu, Y.S. Chen, C.C. Wang, P.J. Tzeng, C.H. Lin, F. Chen, C.H. Lien, and M.-J. Tsai. Low power and high speed bipolar switching with a thin reactive Ti buffer layer in robust HfO_2 based RRAM. In *Electron Devices Meeting, 2008. IEDM 2008. IEEE International*, pages 1 –4, 2008. DOI: 10.1109/IEDM.2008.4796677 Cited on page(s) 14

[105] H.Y. Lee, Y.S. Chen, P.S. Chen, P.Y. Gu, Y.Y. Hsu, S.M. Wang, W.H. Liu, C.H. Tsai, S.S. Sheu, P.C. Chiang, W.P. Lin, C.H. Lin, W.S. Chen, F.T. Chen, C.H. Lien, and M. Tsai. Evidence and solution of over-reset problem for hfox based resistive memory with sub-ns switching speed and high endurance. In *Electron Devices Meeting (IEDM), 2010 IEEE International*, pages 19.7.1 –19.7.4, dec. 2010. DOI: 10.1109/IEDM.2010.5703395 Cited on page(s) 16

[106] Kwang-Jin Lee, Beak-Hyung Cho, Woo-Yeong Cho, Sangbeom Kang, Byung-Gil Choi, Hyung-Rok Oh, Chang-Soo Lee, Hye-Jin Kim, Joon-Min Park, Qi Wang, Mu-Hui Park, Yu-Hwan Ro, Joon-Yong Choi, Ki-Sung Kim, Young-Ran Kim, In-Cheol Shin, Ki-Won Lim, Ho-Keun Cho, Chang-Han Choi, Won-Ryul Chung, Du-Eung Kim, Kwang-Suk Yu, Gi-Tae Jeong, Hong-Sik Jeong, Choong-Keun Kwak, Chang-Hyun Kim, and Kinam Kim. A 90nm 1.8V 512Mb Diode-Switch PRAM with 266MB/s Read Throughput. In *Solid-State Circuits Conference, 2007. ISSCC 2007. Digest of Technical Papers. IEEE International*, pages 472 –616, 2007. DOI: 10.1109/ISSCC.2007.373499 Cited on page(s) 22

[107] Se-Ho Lee, Yeonwoong Jung, and Ritesh Agarwal. Highly scalable non-volatile and ultra-low-power phase-change nanowire memory. *Nature Nanotechnology*, 2:626 – 630, 2007. DOI: 10.1038/nnano.2007.291 Cited on page(s) 21

[108] S.H. Lee, M.S. Kim, G.S. Do, S.G. Kim, H.J. Lee, J.S. Sim, N.G. Park, S.B. Hong, Y.H. Jeon, K.S. Choi, H.C. Park, T.H. Kim, J.U. Lee, H.W. Kim, M.R. Choi, S.Y. Lee, Y.S. Kim, H.J. Kang, J.H. Kim, H.J. Kim, Y.S. Son, B.H. Lee, J.H. Choi, S.C. Kim, J.H. Lee, S.J. Hong, and S.W. Park. Programming disturbance and cell scaling in phase change memory: For up to 16nm based $4F^2$ cell. In *VLSI Technology (VLSIT), 2010 Symposium on*, pages 199 –200, 2010. DOI: 10.1109/VLSIT.2010.5556226 Cited on page(s) 21, 24

[109] Lin Li, Kailiang Lu, B. Rajendran, T.D. Happ, Hsiang-Lan Lung, Chung Lam, and Mansun Chan. Driving device comparison for phase-change memory. *Electron Devices, IEEE Transactions on*, 58(3):664 –671, march 2011. DOI: 10.1109/TED.2010.2100082 Cited on page(s) 22

[110] Albert A. Liddicoat and Michael J. Flynn. Parallel square and cube computations. In *In IEEE 34th Asilomar Confernce on Signals, Systems and Computers*, 2000. DOI: 10.1109/ACSSC.2000.911207 Cited on page(s) 51

[111] K. Lim, J. Chang, T. Mudge, P. Ranganathan, S.K. Reinhardt, and T. Wenisch. Disaggregated Memory for Expansion and Sharing in Blade Servers. pages 267–278, June 2009. DOI: 10.1145/1555815.1555789 Cited on page(s) xi

[112] Jun-Tin Lin, Yi-Bo Liao, Meng-Hsueh Chiang, and Wei-Chou Hsu. Operation of multi-level phase change memory using various programming techniques. In *IC Design and Technology, 2009. ICICDT '09. IEEE International Conference on*, pages 199 –202, May 2009. DOI: 10.1109/ICICDT.2009.5166295 Cited on page(s) 22

[113] L.F. Liu, X. Sun, B. Sun, J.F. Kang, Y. Wang, X.Y. Liu, R.Q. Han, and G.C. Xiong. Current compliance-free resistive switching in nonstoichiometric ceox films for nonvolatile memory application. In *Memory Workshop, 2009. IMW '09. IEEE International*, pages 1 –2, May 2009. DOI: 10.1109/IMW.2009.5090586 Cited on page(s) 14

[114] S. Q. Liu, N. J. Wu, and A. Ignatiev. Electric-pulse-induced reversible resistance change effect in magnetoresistive films. *Applied Physics Letters*, 76(19):2749–2751, 2000. DOI: 10.1063/1.126464 Cited on page(s) 14

[115] Michael Luby and Charles Rackoff. How to construct pseudorandom permutations from pseudorandom functions. *SIAM J. Comput.*, 17(2):373–386, 1988. DOI: 10.1137/0217022 Cited on page(s) 50

[116] M-Systems. *TrueFFS Wear-leveling Mechanism*. http://www.dataio.com/pdf/NAND/MSystems/ TrueFFS_Wear_Leveling_Mec hanism.pdf. Cited on page(s) 44

[117] T.P. Ma and Jin-Ping Han. Why is nonvolatile ferroelectric memory field-effect transistor still elusive? *Electron Device Letters, IEEE*, 23(7):386 –388, July 2002. DOI: 10.1109/LED.2002.1015207 Cited on page(s) 10

[118] S. Madara. The future of cooling high density equipment. *IBM Power and Cooling Technology Symposium*, 2007. Cited on page(s) 66

[119] J.D. Maimon, K.K. Hunt, L. Burcin, and J. Rodgers. Chalcogenide memory arrays: characterization and radiation effects. *Nuclear Science, IEEE Transactions on*, 50(6):1878 – 1884, 2003. DOI: 10.1109/TNS.2003.821377 Cited on page(s) 24

[120] D. Mantegazza, D. Ielmini, E. Varesi, A. Pirovano, and AL Lacaita. Statistical analysis and modeling of programming and retention in PCM arrays. In *IEEE International Electron Devices Meeting, 2007*. DOI: 10.1109/IEDM.2007.4418933 Cited on page(s) 38

[121] Alfred J. Menezes, Paul C. van Oorschot, and Scott A. Vanstone. *Handbook of Applied Cryptography*. 1996. Cited on page(s) 50

[122] Mukut Mitra, Yeonwoong Jung, Daniel S. Gianola, and Ritesh Agarwal. Extremely low drift of resistance and threshold voltage in amorphous phase change nanowire devices. *Applied Physics Letters*, 96(22):222111, 2010. DOI: 10.1063/1.3447941 Cited on page(s) 21

[123] C. Mohan, D. Haderle, B. Lindsay, H. Pirahesh, and P. Schwarz. Aries: A transaction recovery method supporting fine-granularity locking and partial rollbacks using write-ahead logging. *ACM Transactions on Database Systems*, 17(1):94–162, 1992. DOI: 10.1145/128765.128770 Cited on page(s) 88

[124] J. S. Moodera, Lisa R. Kinder, Terrilyn M. Wong, and R. Meservey. Large magnetoresistance at room temperature in ferromagnetic thin film tunnel junctions. *Phys. Rev. Lett.*, 74(16):3273–3276, Apr 1995. DOI: 10.1103/PhysRevLett.74.3273 Cited on page(s) 11

[125] G. E. Moore. Cramming more components onto integrated circuits. *Electronics*, 38(19), 1965. DOI: 10.1109/JPROC.1998.658762 Cited on page(s) 1

[126] R. Muralidhar, R.F. Steimle, M. Sadd, R. Rao, C.T. Swift, E.J. Prinz, J. Yater, L. Grieve, K. Harber, B. Hradsky, S. Straub, B. Acred, W. Paulson, W. Chen, L. Parker, S.G.H. Anderson, M. Rossow, T. Merchant, M. Paransky, T. Huynh, D. Hadad, Ko-Min Chang, and Jr. White, B.E. A 6 V embedded 90 nm silicon nanocrystal nonvolatile memory. In *Electron Devices Meeting, 2003. IEDM '03 Technical Digest. IEEE International*, pages 26.2.1 – 26.2.4, 2003. DOI: 10.1109/IEDM.2003.1269353 Cited on page(s) 9

[127] Computer History Museum. Magnetic Core Memory. http://www.computerhistory.org/revolution/memory-storage/8/253, June. Cited on page(s) 11

[128] H. Nakamoto, D. Yamazaki, T. Yamamoto, H. Kurata, S. Yamada, K. Mukaida, T. Ninomiya, T. Ohkawa, S. Masui, and K. Gotoh. A Passive UHF RFID Tag LSI with 36.6and Current-Mode Demodulator in 0.35 μm FeRAM Technology. In *Solid-State Circuits Conference, 2006. ISSCC 2006. Digest of Technical Papers. IEEE International*, pages 1201 –1210, 2006. DOI: 10.1109/ISSCC.2006.1696166 Cited on page(s) 11

[129] K Nakayama, M Takata, T Kasai, A Kitagawa, and J Akita. Pulse number control of electrical resistance for multi-level storage based on phase change. *Journal of Physics D: Applied Physics*, 40(17):5061, 2007. DOI: 10.1088/0022-3727/40/17/009 Cited on page(s) 22

[130] R. G. Neale, D. L. Nelson, and Gordon E. Moore. Non-volatile, re-programmable, read-mostly memory is here. *Electronics*, pages 56–60, September 1970. Cited on page(s) 17

[131] R.G. Neale and J.A. Aseltine. The application of amorphous materials to computer memories. *Electron Devices, IEEE Transactions on*, 20(2):195 – 205, February 1973. DOI: 10.1109/T-ED.1973.17628 Cited on page(s) 17

[132] T. Nirschl, J.B. Phipp, T.D. Happ, G.W. Burr, B. Rajendran, M.-H. Lee, A. Schrott, M. Yang, M. Breitwisch, C.-F. Chen, E. Joseph, M. Lamorey, R. Cheek, S.-H. Chen, S. Zaidi, S. Raoux, Y.C. Chen, Y. Zhu, R. Bergmann, H.-L. Lung, and C. Lam. Write Strategies for 2 and 4-bit Multi-Level Phase-Change Memory. In *Electron Devices Meeting, 2007. IEDM 2007. IEEE International*, pages 461 –464, 2007. DOI: 10.1109/IEDM.2007.4418973 Cited on page(s) 22, 38

[133] Hiroshi Nozawa and Susumu Kohyama. A Thermionic Electron Emission Model for Charge Retention in SAMOS Structure. *Japanese Journal of Applied Physics*, 21(Part 2, No. 2):L111–L112, 1982. DOI: 10.1143/JJAP.21.L111 Cited on page(s) 4

[134] Library of Congress. Web Arching FAQ page from Library of Congress. http://www.loc.gov/webarchiving/faq.html, June 2009. Cited on page(s) 1

[135] G.H. Oh, Y.L. Park, J.I. Lee, D.H. Im, J.S. Bae, D.H. Kim, D.H. Ahn, H. Horii, S.O. Park, H.S. Yoon, I.S. Park, Y.S. Ko, U-In Chung, and J.T. Moon. Parallel multi-confined (PMC) cell technology for high density MLC PRAM. In *VLSI Technology, 2009 Symposium on*, pages 220 –221, 2009. Cited on page(s) 22

[136] J.H. Oh, J.H. Park, Y.S. Lim, H.S. Lim, Y.T. Oh, J.S. Kim, J.M. Shin, Y.J. Song, K.C. Ryoo, D.W. Lim, S.S. Park, J.I. Kim, J.H. Kim, J. Yu, F. Yeung, C.W. Jeong, J.H. Kong, D.H. Kang, G.H. Koh, G.T. Jeong, H.S. Jeong, and Kinam Kim. Full integration of highly manufacturable 512mb pram based on 90nm technology. In *Electron Devices Meeting, 2006. IEDM '06. International*, pages 1 –4, 2006. DOI: 10.1109/IEDM.2006.346905 Cited on page(s) 22

[137] T. Ohta. Phase Change Optical Memory Promotes the DVD Optical Disk. *Journal of Optoelectronics and Advanced Materials*, 3(3):609 – 626, 2001. Cited on page(s) 20

[138] Wataru Otsuka, Koji Miyata, Makoto Kitagawa, Keiichi Tsutsui, Tomohito Tsushima, Hiroshi Yoshihara, Tomohiro Namise, Yasuhiro Terao, and Kentaro Ogata. A 4mb conductive-bridge resistive memory with 2.3gb/s read-throughput and 216mb/s program-throughput. In *Solid-State Circuits Conference Digest of Technical Papers (ISSCC), 2011 IEEE International*, pages 210 –211, feb. 2011. DOI: 10.1109/ISSCC.2011.5746286 Cited on page(s) 16

[139] S. Ovshinsky and H. Fritzsche. Reversible structural transformations in amorphous semiconductors for memory and logic. *Metallurgical and Materials Transactions B*, 2:641–645, 1971. 10.1007/BF02662715. DOI: 10.1007/BF02662715 Cited on page(s) 17

[140] Stanford R. Ovshinsky. Reversible electrical switching phenomena in disordered structures. *Phys. Rev. Lett.*, 21(20):1450–1453, Nov 1968. DOI: 10.1103/PhysRevLett.21.1450 Cited on page(s) 17

[141] Ki-Tae Park, Ohsuk Kwon, Sangyong Yoon, Myung-Hoon Choi, In-Mo Kim, Bo-Geun Kim, Min-Seok Kim, Yoon-Hee Choi, Seung-Hwan Shin, Youngson Song, Joo-Yong Park, Jae-Eun Lee, Chang-Gyu Eun, Ho-Chul Lee, Hyeong-Jun Kim, Jun-Hee Lee, Jong-Young Kim, Tae-Min Kweon, Hyun-Jun Yoon, Taehyun Kim, Dong-Kyo Shim, Jongsun Sel, Ji-Yeon Shin, Pansuk Kwak, Jin-Man Han, Keon-Soo Kim, Sungsoo Lee, Young-Ho Lim, and Tae-Sung Jung. A 7MB/s 64Gb 3-Bit/Cell DDR NAND Flash Memory in 20nm-Node Technology. In *Solid-State Circuits Conference, 2011. Digest of Technical Papers.*, 2011. DOI: 10.1109/ISSCC.2011.5746287 Cited on page(s) 8, 9

[142] Stuart S. P. Parkin, Christian Kaiser, Alex Panchula, Philip M. Rice, Brian Hughes, Mahesh Samant, and See-Hun Yang. Giant tunnelling magnetoresistance at room temperature with MgO (100) tunnel barriers. *Nature Materials*, 3, 2004. DOI: 10.1038/nmat1256 Cited on page(s) 11

[143] P. Pavan, R. Bez, P. Olivo, and E. Zanoni. Flash memory cells-an overview. *Proceedings of the IEEE*, 85(8):1248 –1271, August 1997. DOI: 10.1109/5.622505 Cited on page(s) 6

[144] Ed Jr. Pegg. "Lebombo Bone." From MathWorld–A Wolfram Web Resource, created by Eric W. Weisstein. http://mathworld.wolfram.com/LebomboBone.html, June 2009. Cited on page(s) 1

[145] A. Pirovano. Emerging Non-Volatile Memories. IEDM Short course, June 2006. Cited on page(s) 8

[146] A. Pirovano, A. Redaelli, F. Pellizzer, F. Ottogalli, M. Tosi, D. Ielmini, A.L. Lacaita, and R. Bez. Reliability study of phase-change nonvolatile memories. *Device and Materials Reliability, IEEE Transactions on*, 4(3):422 – 427, 2004. DOI: 10.1109/TDMR.2004.836724 Cited on page(s) 23

[147] A. Pohm, C. Sie, R. Uttecht, V. Kao, and O. Agrawal. Chalcogenide glass bistable resistivity (ovonic) memories. *Magnetics, IEEE Transactions on*, 6(3):592, September 1970. DOI: 10.1109/TMAG.1970.1066920 Cited on page(s) 17

[148] A.V. Pohm, J.S.T. Huang, J.M. Daughton, D.R. Krahn, and V. Mehra. The design of a one megabit non-volatile M-R memory chip using 1.5 ×5 μm cells. *Magnetics, IEEE Transactions on*, 24(6):3117 –3119, November 1988. DOI: 10.1109/20.92353 Cited on page(s) 11

[149] V. Prabhakaran, T.L. Rodeheffer, and L. Zhou. Transactional Flash. In *USENIX Symposium on Operating Systems Design and Implementation (OSDI)*, pages 147–160, December 2008. Cited on page(s) 84

[150] I L Prejbeanu, M Kerekes, R C Sousa, H Sibuet, O Redon, B Dieny, and J P Nozires. Thermally assisted mram. *Journal of Physics: Condensed Matter*, 19(16):165218, 2007. DOI: 10.1088/0953-8984/19/16/165218 Cited on page(s) 13

[151] S. Privitera, C. Bongiorno, E. Rimini, and R. Zonca. Crystal nucleation and growth processes in $Ge_2Sb_2Te_5$. *Applied Physics Letters*, 84(22):4448–4450, 2004. DOI: 10.1063/1.1759063 Cited on page(s) 19

[152] E. W. Pugh, D. L. Critchlow, R. A. Henle, and L. A. Russell. Solid State Memory Development in IBM. *IBM Journal of Research and Development*, 25(5):585 –602, 1981. DOI: 10.1147/rd.255.0585 Cited on page(s) 11

[153] Charles F. Pulvari. An Electrostatically Induced Permanent Memory. *Journal of Applied Physics*, 22, 1951. DOI: 10.1063/1.1700098 Cited on page(s) 9

[154] M.K. Qureshi, J. Karidis, M. Franceschini, V. Srinivasan, L. Lastras, and B. Abali. Enhancing Lifetime and Security of Phase Change Memories via Start-Gap Wear Leveling. In *Proceedings of the International Symposium on Microarchitecture (MICRO)*, pages 14–23, December 2009. DOI: 10.1145/1669112.1669117 Cited on page(s) 43, 44, 45, 50, 55, 56, 65, 82

[155] Moinuddin Qureshi et al. Scalable high performance main memory system using phase-change memory technology. In *ISCA-36*, 2009. DOI: 10.1145/1555815.1555760 Cited on page(s) 29, 31, 32

[156] Moinuddin K. Qureshi et al. Practical and secure pcm systems by online detection of malicious write streams. In *HPCA-2011*. DOI: 10.1109/HPCA.2011.5749753 Cited on page(s) 55, 60, 61, 64

[157] Moinuddin K. Qureshi et al. Improving read performance of phase change memories via write cancellation and write pausing. In *HPCA-16*, 2010. DOI: 10.1109/HPCA.2010.5416645 Cited on page(s) 35, 37, 38, 65

[158] Moinuddin K. Qureshi, Vijayalakshmi Srinivasan, and Jude A. Rivers. Scalable high performance main memory system using phase-change memory technology. In *Proceedings of the 36th annual international symposium on Computer architecture*, ISCA '09, pages 24–33, New York, NY, USA, 2009. ACM. DOI: 10.1145/1555815.1555760 Cited on page(s) 3

[159] B. Rajendran, M. Breitwisch, Ming-Hsiu Lee, G.W. Burr, Yen-Hao Shih, R. Cheek, A. Schrott, Chieh-Fang Chen, E. Joseph, R. Dasaka, H.-L. Lung, and Chung Lam. Dynamic Resistance - A Metric for Variability Characterization of Phase-Change Memory. *Electron*

Device Letters, IEEE, 30(2):126 –129, 2009. DOI: 10.1109/LED.2008.2010004   Cited on page(s) 21

[160] B. Rajendran, J. Karidis, M.-H. Lee, M. Breitwisch, G.W. Burr, Y.-H. Shih, R. Cheek, A. Schrott, H.-L. Lung, and C. Lam. Analytical model for reset operation of phase change memory. In *Electron Devices Meeting, 2008. IEDM 2008. IEEE International*, pages 1 –4, 2008. DOI: 10.1109/IEDM.2008.4796748 Cited on page(s) 21

[161] B. Rajendran, M.-H. Lee, M. Breitwisch, G.W. Burr, Y.-H. Shih, R. Cheek, A. Schrott, C.-F. Chen, M. Lamorey, E. Joseph, Y. Zhu, R. Dasaka, P.L. Flaitz, F.H. Baumann, H.-L. Lung, and C. Lam. On the dynamic resistance and reliability of phase change memory. In *VLSI Technology, 2008 Symposium on*, pages 96 –97, 2008. DOI: 10.1109/VLSIT.2008.4588576 Cited on page(s) 19, 24

[162] B. Rajendran, R.S. Shenoy, D.J. Witte, N.S. Chokshi, R.L. De Leon, G.S. Tompa, and R. Fabian. Low Thermal Budget Processing for Sequential 3-D IC Fabrication. *Electron Devices, IEEE Transactions on*, 54(4):707 –714, 2007. DOI: 10.1109/TED.2007.891300 Cited on page(s) 4, 5

[163] Feng Rao, Zhitang Song, Min Zhong, Liangcai Wu, Gaoming Feng, Bo Liu, Songlin Feng, and Bomy Chen. Multilevel data storage characteristics of phase change memory cell with doublelayer chalcogenide films ($ge_2sb_2te_5$ and sb_2te_3). *Japanese Journal of Applied Physics*, 46(2):L25–L27, 2007. DOI: 10.1143/JJAP.46.L25 Cited on page(s) 22

[164] S. Raoux and M. Wuttig. *Phase Change Materials*. Springer-Verlag, Berlin-Heidelberg, 2009. Cited on page(s) 21

[165] Simone Raoux, Jean L. Jordan-Sweet, and Andrew J. Kellock. Crystallization properties of ultrathin phase change films. *Journal of Applied Physics*, 103(11):114310, 2008. DOI: 10.1063/1.2938076 Cited on page(s) 21

[166] Simone Raoux, Charles T. Rettner, Jean L. Jordan-Sweet, Andrew J. Kellock, Teya Topuria, Philip M. Rice, and Dolores C. Miller. Direct observation of amorphous to crystalline phase transitions in nanoparticle arrays of phase change materials. *Journal of Applied Physics*, 102(9):094305, 2007. DOI: 10.1063/1.2801000 Cited on page(s) 18, 21, 23

[167] A. Redaelli, A. Pirovano, A. Benvenuti, and A. L. Lacaita. Threshold switching and phase transition numerical models for phase change memory simulations. *Journal of Applied Physics*, 103(11):111101, 2008. DOI: 10.1063/1.2931951 Cited on page(s) 19

[168] A. Redaelli, A. Pirovano, A. Locatelli, and F. Pellizzer. Numerical implementation of low field resistance drift for phase change memory simulations. In *Non-Volatile Semiconductor Memory Workshop, 2008 and 2008 International Conference on Memory Technology and Design.*

NVSMW/ICMTD 2008. Joint, pages 39 –42, May 2008. DOI: 10.1109/NVSMW.2008.17 Cited on page(s) 23

[169] J. Reifenberg, E. Pop, A. Gibby, S. Wong, and K. Goodson. Multiphysics Modeling and Impact of Thermal Boundary Resistance in Phase Change Memory Devices. In *Thermal and Thermomechanical Phenomena in Electronics Systems, 2006. ITHERM '06. The Tenth Intersociety Conference on*, pages 106 –113, 30 2006-june 2 2006. DOI: 10.1109/ITHERM.2006.1645329 Cited on page(s) 24

[170] Samsung corporation. K9XXG08XXM flash memory specification. http://www.samsung.com/global/system/business/semiconductor/product/2007/6/11/NANDFlash/SLC_LargeBlock/8Gbit/K9F8G08U0M/ds_k9f8g08x0m_rev10.pdf. Cited on page(s) 81

[171] Joy Sarkar and Bob Gleixner. Evolution of phase change memory characteristics with operating cycles: Electrical characterization and physical modeling. *Applied Physics Letters*, 91(23):233506, 2007. DOI: 10.1063/1.2821845 Cited on page(s) 19

[172] A. Sawa, T. Fujii, M. Kawasaki, and Y. Tokura. Hysteretic current–voltage characteristics and resistance switching at a rectifying $Ti/Pr_{0.7}Ca_{0.3}MnO_3$ interface. *Applied Physics Letters*, 85(18):4073–4075, 2004. DOI: 10.1063/1.1812580 Cited on page(s) 15

[173] Akihito Sawa. Resistive switching in transition metal oxides. *Materials Today*, 11(6):28 – 36, 2008. DOI: 10.1016/S1369-7021(08)70119-6 Cited on page(s) 14, 15

[174] Akihito Sawa, Takeshi Fujii, Masashi Kawasaki, and Yoshinori Tokura. Colossal Electro-Resistance Memory Effect at Metal/La_2CuO_4 Interfaces. *Japanese Journal of Applied Physics*, 44(40):L1241–L1243, 2005. DOI: 10.1143/JJAP.44.L1241 Cited on page(s) 15

[175] Stuart Schechter, Gabriel H. Loh, Karin Strauss, and Doug Burger. Use ECP, not ECC, for hard failures in resistive memories. In *ISCA-37*, 2010. DOI: 10.1145/1816038.1815980 Cited on page(s) 69, 72, 73

[176] R. Sears and E. Brewer. Statis: Flexible Transactional Storage. In *USENIX Symposium on Operating Systems Design and Implementation (OSDI)*, pages 29–44, November 2006. Cited on page(s) 89

[177] T. Sekiguchi, K. Ono, A. Kotabe, and Y. Yanagawa. 1-Tbyte/s 1-Gbit DRAM Architecture Using 3-D Interconnect for High-Throughput Computing. *Solid-State Circuits, IEEE Journal of*, 46(4):828 –837, 2011. DOI: 10.1109/JSSC.2011.2109630 Cited on page(s) 3

[178] M. Seltzer and M. Olsen. LIBTP: Portable, Modular Transactions for UNIX. In *Proceedings of the Winter Usenix Conference*, pages 9–25, January 1992. Cited on page(s) 89

[179] Nak Hee Seong, Dong Hyuk Woo, and Hsien-Hsin S. Lee. Security refresh: Prevent malicious wear-out and increase durability for phase-change memory with dynamically randomized address mapping. In *ISCA-37*, 2010. DOI: 10.1145/1816038.1816014 Cited on page(s) 56, 57, 67

[180] Nak Hee Seongy et al. SAFER: Stuck-At-Fault Error Recovery for Memories. In *MICRO-43*, 2010. DOI: 10.1109/MICRO.2010.46 Cited on page(s) 69, 74, 75

[181] G. Servalli. A 45nm generation phase change memory technology. In *Electron Devices Meeting (IEDM), 2009 IEEE International*, pages 1 –4, 2009. DOI: 10.1109/IEDM.2009.5424409 Cited on page(s) 16, 22, 23, 24

[182] Andre Seznec. Towards phase change memory as a secure main memory. Technical report, INRIA, November 2009. Cited on page(s) 56

[183] Andre Seznec. A phase change memory as a secure main memory. *IEEE Computer Architecture Letters*, Feb 2010. DOI: 10.1109/L-CA.2010.2 Cited on page(s) 57

[184] R. Shanks and C. Davis. A 1024-bit nonvolatile 15ns bipolar read-write memory. In *Solid-State Circuits Conference. Digest of Technical Papers. 1978 IEEE International*, volume XXI, pages 112 – 113, February 1978. DOI: 10.1109/ISSCC.1978.1155755 Cited on page(s) 17

[185] J. J. Shedletsky. Error correction by alternate-data retry. *IEEE Trans. Comput.*, 27, February 1978. DOI: 10.1109/TC.1978.1675044 Cited on page(s) 74

[186] Shyh-Shyuan Sheu, Meng-Fan Chang, Ku-Feng Lin, Che-Wei Wu, Yu-Sheng Chen, Pi-Feng Chiu, Chia-Chen Kuo, Yih-Shan Yang, Pei-Chia Chiang, Wen-Pin Lin, Che-He Lin, Heng-Yuan Lee, Pei-Yi Gu, Sum-Min Wang, Frederick T. Chen, Keng-Li Su, Chen-Hsin Lien, Kuo-Hsing Cheng, Hsin-Tun Wu, Tzu-Kun Ku, Ming-Jer Kao, and Ming-Jinn Tsai. A 4mb embedded slc resistive-ram macro with 7.2ns read-write random-access time and 160ns mlc-access capability. In *Solid-State Circuits Conference Digest of Technical Papers (ISSCC), 2011 IEEE International*, pages 200 –202, feb. 2011. DOI: 10.1109/ISSCC.2011.5746281 Cited on page(s) 16

[187] H. Shiga, D. Takashima, S. Shiratake, K. Hoya, T. Miyakawa, R. Ogiwara, R. Fukuda, R. Tak-izawa, K. Hatsuda, F. Matsuoka, Y. Nagadomi, D. Hashimoto, H. Nishimura, T. Hioka, S. Doumae, S. Shimizu, M. Kawano, T. Taguchi, Y. Watanabe, S. Fujii, T. Ozaki, H. Kanaya, Y. Kumura, Y. Shimojo, Y. Yamada, Y. Minami, S. Shuto, K. Yamakawa, S. Yamazaki, I. Ku-nishima, T. Hamamoto, A. Nitayama, and T. Furuyama. A 1.6GB/s DDR2 128Mb chain FeRAM with scalable octal bitline and sensing schemes. In *Solid-State Circuits Conference - Digest of Technical Papers, 2009. ISSCC 2009. IEEE International*, pages 464 –465,465a, 2009. DOI: 10.1109/ISSCC.2009.4977509 Cited on page(s) 11, 16

[188] Y.H. Shih, J.Y. Wu, B. Rajendran, M.H. Lee, R. Cheek, M. Lamorey, M. Breitwisch, Y. Zhu, E.K. Lai, C.F. Chen, E. Stinzianni, A. Schrott, E. Joseph, R. Dasaka, S. Raoux, H.L. Lung, and C. Lam. Mechanisms of retention loss in $Ge_2Sb_2Te_5$-based Phase-Change Memory. In *Electron Devices Meeting, 2008. IEDM 2008. IEEE International*, pages 1 –4, 2008. DOI: 10.1109/IEDM.2008.4796653 Cited on page(s) 23

[189] Jungho Shin, Insung Kim, Kuyyadi P. Biju, Minseok Jo, Jubong Park, Joonmyoung Lee, Seungjae Jung, Wootae Lee, Seonghyun Kim, Sangsu Park, and Hyunsang Hwang. TiO_2-based metal-insulator-metal selection device for bipolar resistive random access memory cross-point application. *Journal of Applied Physics*, 109(3):033712, 2011. DOI: 10.1063/1.3544205 Cited on page(s) 14

[190] J.C. Slonczewski. Current-driven excitation of magnetic multilayers. *Journal of Magnetism and Magnetic Materials*, 159(1-2):L1 – L7, 1996. DOI: 10.1016/0304-8853(96)00062-5 Cited on page(s) 13

[191] C. Smullen, J. Coffman, and S. Gurumurthi. Accelerating Enterprise Solid-State Disks with Non-Volatile Merge Caching. In *Proceedings of the International Green Computing Conference (IGCC)*, August 2010. DOI: 10.1109/GREENCOMP.2010.5598310 Cited on page(s) 83, 84

[192] C.W. Smullen, V. Mohan, A. Nigam, S. Gurumurthi, and M.R. Stan. Relaxing Non-Volatility for Fast and Energy-Efficient STT-RAM Caches. In *Proceedings of the International Symposium on High Performance Computer Architecture (HPCA)*, pages 50–61, February 2011. DOI: 10.1109/HPCA.2011.5749716 Cited on page(s) 5, 91, 92

[193] T. Sugizaki, M. Kobayashi, M. Ishidao, H. Minakata, M. Yamaguchi, Y. Tamura, Y. Sugiyama, T. Nakanishi, and H. Tanaka. Novel multi-bit SONOS type flash memory using a high-k charge trapping layer. In *VLSI Technology, 2003. Digest of Technical Papers. 2003 Symposium on*, pages 27 – 28, 2003. DOI: 10.1109/VLSIT.2003.1221069 Cited on page(s) 9

[194] B. Sun, Y. X. Liu, L. F. Liu, N. Xu, Y. Wang, X. Y. Liu, R. Q. Han, and J. F. Kang. Highly uniform resistive switching characteristics of $TiN/ZrO_2/Pt$ memory devices. *Journal of Applied Physics*, 105(6):061630, 2009. DOI: 10.1063/1.3055414 Cited on page(s) 14

[195] Guangyu Sun, Xiangyu Dong, Yuan Xie, Jian Li, and Yiran Chen. A novel architecture of the 3d stacked mram l2 cache for cmps. In *IEEE 15th International Symposium on High Performance Computer Architecture*, 2009. DOI: 10.1109/HPCA.2009.4798259 Cited on page(s) 36

[196] Zhimei Sun, Jian Zhou, Andreas Blomqvist, Börje Johansson, and Rajeev Ahuja. Formation of large voids in the amorphous phase-change memory $ge_2sb_2te_5$ alloy. *Phys. Rev. Lett.*, 102(7):075504, Feb 2009. DOI: 10.1103/PhysRevLett.102.075504 Cited on page(s) 24

[197] D. Takashima and I. Kunishima. High-density chain ferroelectric random access memory (chain FRAM) . *Solid-State Circuits, IEEE Journal of*, 33(5):787 –792, May 1998. DOI: 10.1109/4.668994 Cited on page(s) 10

[198] D.D. Tang, P.K. Wang, V.S. Speriosu, S. Le, and K.K. Kung. Spin-valve RAM cell. *Magnetics, IEEE Transactions on*, 31(6):3206 –3208, November 1995. DOI: 10.1109/20.490329 Cited on page(s) 11

[199] K. Tsuchida, T. Inaba, K. Fujita, Y. Ueda, T. Shimizu, Y. Asao, T. Kajiyama, M. Iwayama, K. Sugiura, S. Ikegawa, T. Kishi, T. Kai, M. Amano, N. Shimomura, H. Yoda, and Y. Watanabe. A 64Mb MRAM with clamped-reference and adequate-reference schemes. In *Solid-State Circuits Conference Digest of Technical Papers (ISSCC), 2010 IEEE International*, pages 258 –259, 2010. DOI: 10.1109/ISSCC.2010.5433948 Cited on page(s) 13

[200] T Tsuruoka, K Terabe, T Hasegawa, and M Aono. Forming and switching mechanisms of a cation-migration-based oxide resistive memory. *Nanotechnology*, 21(42):425205, 2010. DOI: 10.1088/0957-4484/21/42/425205 Cited on page(s) 15

[201] Tze-chiang Chen. Innovation in solid state devices for exascale computing. In *VLSI Technology Systems and Applications (VLSI-TSA), 2010 International Symposium on*, pages 2 –5, 2010. DOI: 10.1109/VTSA.2010.5488971 Cited on page(s) 1

[202] L. van Pieterson, M. van Schijndel, J. C. N. Rijpers, and M. Kaiser. Te-free, sb-based phase-change materials for high-speed rewritable optical recording. *Applied Physics Letters*, 83(7):1373–1375, 2003. DOI: 10.1063/1.1604172 Cited on page(s) 19

[203] S. Venkataraman, N. Tolia, P. Ranganathan, and R.H. Campbell. Consistent and Durable Data Structures for Non-Volatile Byte-Addressable Memory. In *Proceedings of the USENIX Conference on File and Storage Technonologies (FAST)*, February 2011. Cited on page(s) 89

[204] H. Volos, A.J. Tack, and M.M. Swift. Mnemosyne: Lightweight Persistent Memory. In *Proceedings of the International Conference on Architectural Support for Programming Languages and Operating Systems (ASPLOS)*, pages 91–104, March 2011. DOI: 10.1145/1961296.1950379 Cited on page(s) 89

[205] Sheng-Yu Wang, Chin-Wen Huang, Dai-Ying Lee, Tseung-Yuen Tseng, and Ting-Chang Chang. Multilevel resistive switching in Ti/Cu$_x$O/Pt memory devices. *Journal of Applied Physics*, 108(11):114110, 2010. DOI: 10.1063/1.3518514 Cited on page(s) 14

[206] X.Q. Wei, L.P. Shi, R. Walia, T.C. Chong, R. Zhao, X.S. Miao, and B.S. Quek. Hspice macromodel of pcram for binary and multilevel storage. *Electron Devices, IEEE Transactions on*, 53(1):56 – 62, jan. 2006. DOI: 10.1109/TED.2005.860645 Cited on page(s) 25

[207] X.Q. Wei, L.P. Shi, R. Walia, T.C. Chong, R. Zhao, X.S. Miao, and B.S. Quek. Hspice macro-model of pcram for binary and multilevel storage. *APPLIED PHYSICS A: MATERIALS SCIENCE AND PROCESSING*, 102(4):765–783, jan. 2011. DOI: 10.1109/TED.2005.860645 Cited on page(s) 15

[208] Z. Wei, Y. Kanzawa, K. Arita, Y. Katoh, K. Kawai, S. Muraoka, S. Mitani, S. Fujii, K. Katayama, M. Iijima, T. Mikawa, T. Ninomiya, R. Miyanaga, Y. Kawashima, K. Tsuji, A. Himeno, T. Okada, R. Azuma, K. Shimakawa, H. Sugaya, T. Takagi, R. Yasuhara, K. Horiba, H. Kumigashira, and M. Oshima. Highly reliable TaOx ReRAM and direct evidence of redox reaction mechanism. In *Electron Devices Meeting, 2008. IEDM 2008. IEEE International*, pages 1 –4, 2008. DOI: 10.1109/IEDM.2008.4796676 Cited on page(s) 14

[209] H.P. Wong, S. Raoux, S. Kim, J. Liang, J.P. Reifenberg, B. Rajendran, M. Asheghi, and K.E. Goodson. Phase change memory. *Proceedings of the IEEE*, 98(12):2201 –2227, 2010. DOI: 10.1109/JPROC.2010.2070050 Cited on page(s) 17, 18

[210] D. C. Worledge. Single-domain model for toggle MRAM. *IBM Journal of Research and Development*, 50(1):69 –79, 2006. DOI: 10.1147/rd.501.0069 Cited on page(s) 12

[211] D.C. Worledge, G. Hu, P.L. Trouilloud, D.W. Abraham, S. Brown, M.C. Gaidis, J. Nowak, E.J. O'Sullivan, R.P. Robertazzi, J.Z. Sun, and W.J. Gallagher. Switching distributions and write reliability of perpendicular spin torque MRAM. In *Electron Devices Meeting (IEDM), 2010 IEEE International*, pages 12.5.1 –12.5.4, 2010. DOI: 10.1109/IEDM.2010.5703349 Cited on page(s) 13

[212] Wei Xu and Tong Zhang. Using time-aware memory sensing to address resistance drift issue in multi-level phase change memory. In *Quality Electronic Design (ISQED), 2010 11th International Symposium on*, pages 356 –361, 2010. DOI: 10.1109/ISQED.2010.5450549 Cited on page(s) 23

[213] J Joshua Yang, Feng Miao, Matthew D Pickett, Douglas A A Ohlberg, Duncan R Stewart, Chun Ning Lau, and R Stanley Williams. The mechanism of electroforming of metal oxide memristive switches. *Nanotechnology*, 20(21):215201, 2009. DOI: 10.1088/0957-4484/20/21/215201 Cited on page(s) 15

[214] Doe Hyun Yoon, Naveen Muralimanohar, Jichuan Chang, Parthasarathy Ranganathan, Norman P. Jouppi, and Mattan Erez. FREE-p: Protecting non-volatile memory against both hard and soft errors. In *HPCA-2011*. DOI: 10.1109/HPCA.2011.5749752 Cited on page(s) 69, 76

[215] Lijie Zhang, Minghao Zhu, Ru Huang, Dejin Gao, Yongbian Kuang, Congyin Shi, and Yangyuan Wang. Forming-Less Unipolar TaOx-Based RRAM with Large CC-Independence Range for High Density Memory Applications. *ECS Transactions*, 27(1):3–8, 2010. DOI: 10.1149/1.3360587 Cited on page(s) 14

[216] M. Zhong, Z.T. Song, B. Liu, L.Y. Wang, and S.L. Feng. Switching reliability improvement of phase change memory with nanoscale damascene structure by $Ge_2Sb_2Te_5$ CMP process. *Electronics Letters*, 44(4):322 –323, 14 2008. DOI: 10.1049/el:20082906 Cited on page(s) 24

[217] Guofu Zhou, Herman J. Borg, J. C. N. Rijpers, Martijn H. R. Lankhorst, and J. J. L. Horikx. Crystallization behavior of phase-change materials: comparison between nucleation- and growth-dominated crystallization.   volume 4090, pages 108–115. SPIE, 2000. DOI: 10.1109/ODS.2000.847985 Cited on page(s) 19

[218] Ping Zhou, Bo Zhao, Jun Yang, and Youtao Zhang. A durable and energy efficient main memory using phase change memory technology. In *ISCA-36*, 2009. DOI: 10.1145/1555815.1555759 Cited on page(s) 29, 30, 31, 56

Authors' Biography

MOINUDDIN QURESHI

Dr. Moinuddin Qureshi is an Associate Professor at Georgia Institute of Technology. His research interest includes computer architecture, memory system design, and leveraging emerging technology for scalable and efficient systems. He was a Research Staff Member at IBM T.J. Watson Research Center from 2007 to 2011, where he contributed to caching algorithms of Power 7 processor and conducted research studies on emerging non-volatile memory technologies. He received his Ph.D. (2007) and M.S. (2003) from the University of Texas at Austin, and BE (2000) from Mumbai University. He has published more than a dozen papers in flagship architecture conferences and holds five US patents.

SUDHANVA GURUMURTHI

Dr. Sudhanva Gurumurthi is an Associate Professor in the Computer Science Department at the University of Virginia. He received a BE degree from the College of Engineering Guindy, Chennai, India in 2000 and his Ph.D. from Penn State in 2005, both in the field of Computer Science and Engineering. Sudhanva's research interests include memory and storage systems, processor fault tolerance, and data center architecture. He has served on the program and organizing committees of several top computer architecture and systems conferences including ISCA, ASPLOS, HPCA, FAST, and SIGMETRICS. He has been an Associate Editor-in-Chief for IEEE Computer Architecture Letters (CAL) and currently serves as an Associate Editor. Sudhanva has held research positions at IBM Research and Intel and has served as a faculty consultant for Intel. Sudhanva is a recipient of the NSF CAREER Award and has received several research awards from NSF, Intel, Google, and HP. He is a Senior Member of the IEEE and the ACM.

BIPIN RAJENDRAN

Dr. Bipin Rajendran is a Master Inventor and Research Staff Member at IBM T.J. Watson Research Center, engaged in exploratory research on non-volatile memories and neuromorphic computation. He has contributed to works that led to the most advanced multi-level demonstration in PCM (Nirschl et al, IEDM '07), universal metrics for reliability characterization of PCM (Rajendran et al, VLSI Technology Symposium '08), analytical model for PCM operation (Rajendran et al, IEDM '08) and PCM data retention models (Y.H Shih et al, IEDM '08). He has published more than 30 papers in peer reviewed journals and conferences and holds 20 US patents. He has served as a member of the Emerging Research Devices Working Group of the International Technology Roadmap for Semiconductors (ITRS) in 2010. He received a B.Tech degree (2000) from Indian Institute of Technology, Kharagpur and M.S (2003) and Ph.D (2006) in Electrical Engineering from Stanford University.

Printed in the United States
by Baker & Taylor Publisher Services